AF388955

New Challenges in Water Systems

New Challenges in Water Systems

Editors

Helena M. Ramos
Armando Carravetta
Aonghus Mc Nabola

MDPI • Basel • Beijing • Wuhan • Barcelona • Belgrade • Manchester • Tokyo • Cluj • Tianjin

Editors
Helena M. Ramos
University of Lisbon
Portugal

Armando Carravetta
University Federico II of Naples
Italy

Aonghus Mc Nabola
Trinity College
Ireland

Editorial Office
MDPI
St. Alban-Anlage 66
4052 Basel, Switzerland

This is a reprint of articles from the Special Issue published online in the open access journal *Water* (ISSN 2073-4441) (available at: https://www.mdpi.com/journal/water/special_issues/water_system_challenges).

For citation purposes, cite each article independently as indicated on the article page online and as indicated below:

LastName, A.A.; LastName, B.B.; LastName, C.C. Article Title. *Journal Name* **Year**, *Article Number*, Page Range.

ISBN 978-3-03943-276-9 (Hbk)
ISBN 978-3-03943-277-6 (PDF)

Cover image courtesy of Helena M. Ramos.

© 2020 by the authors. Articles in this book are Open Access and distributed under the Creative Commons Attribution (CC BY) license, which allows users to download, copy and build upon published articles, as long as the author and publisher are properly credited, which ensures maximum dissemination and a wider impact of our publications.

The book as a whole is distributed by MDPI under the terms and conditions of the Creative Commons license CC BY-NC-ND.

Contents

About the Editors

Helena M. Ramos is a Professor at Instituto Superior Técnico—IST (the Engineering Faculty from University of Lisbon) and a member of CERIS in Civil Engineering Department. She has participated in 15 national projects and in 12 International Scientific projects and has been the supervisor of 20 PhD students, 45 MSc theses, and 10 post-graduates/docs. She has more than 5382 citations, index h 40 and i10 119. She has been a Member of various Editorial Teams and Reviewer of different scientific journals. She has several publications, more than 150 in International Journals with referees, more than 180 in Conferences, 22 book chapters, 3 books: in Small Hydropower Plants 2000 and Pump as Turbines 2018, Bombas operando como Turbinas, among others documents of technical and scientific disclosure. She has received 3 International Scientific recognitions. She is an expert in transients, hydropower, pumped storage and hybrid renewable solutions, water –energy nexus, leakage detection, hydrodynamics, and hemodynamics. For more detail: http://scholar.google.pt/citations?sortby=pubdate&hl=pt-PT&user=9jTHk6oAAAAJ; http://orcid.org/0000-0002-9028-9711.

Armando Carravetta is a Full Professor in Hydraulics at the Department of Civil, Architecture and Environmental Engineering, University of Naples, Federico II, IT. He undertakes research in technical innovation for water systems, energy efficiency and resilience of pumping systems, energy recovery by PAT technology, and fluid dynamics of slurry flows. He represents the Italian Association of Pump Manufacturers in the Europump Lot 28 Working Group for the implementation of the European standards on Ecodesign. He is one of the partners in the ongoing EU Interreg project, Reducing Energy Dependency in Atlantic Area Water Networks (REDAWN).

Aonghus Mc Nabola is a Professor in Energy and the Environment at the Department of Civil, Structural and Environmental Engineering, TCD. His research interests lie in the field of environmental fluid dynamics, where he has applied this expertise in the air pollution, energy efficiency, and/or water services sectors. He has been active in this field of research for over 15 years. Prof. McNabola currently leads a group of 8 PhD students and 3 postdoctoral researchers and is involved, primarily, as the lead partner and principal investigator, in a number of national and international collaborative projects, funded by Horizon 2020, INTERREG ERDF, SEAI, and the EPA. Prof. McNabola has a h-index of 25 and an i-10 index of 45. He has generated funding of over 5.25 million from national and primarily EU sources in over 20 funded research projects since 2008.

Preface to "New Challenges in Water Systems"

This new era requires new thinking and focused resolve, through the identification of the biggest challenges and concerns in the water sector, to support the development of new design solutions and analyses. The decision on future directions is to look closely at what the key issues are, such as system efficiency, smart water grids, advanced simulations and analyses, loss control and gain opportunities, innovative integrated solutions, and water–energy management, which researchers and engineers must address today toward future challenges, research directions, and applications.

This Special Issue aims to provide an investigation and engineering opportunity, where scientists, researchers, and experts can submit their novel developments, new design solutions, innovative approaches in several fields of hydraulics, techniques, methods, and analyses in order to respond to the new challenges in the water sector.

Helena M. Ramos, Armando Carravetta, Aonghus Mc Nabola
Editors

Editorial

New Challenges in Water Systems

Helena M. Ramos [1],*, Armando Carravetta [2] and Aonghus Mc Nabola [3]

[1] Department of Civil Engineering and Architecture, CERIS, Instituto Superior Técnico, University of Lisbon, 1049-001 Lisbon, Portugal
[2] Department of Civil, Architecture and Environmental Engineering, University Federico II of Naples, 80125 Naples, Italy; armando.carravetta@unina.it
[3] Department of Civil, Structural and Environmental Engineering, Trinity College Dublin, D02 PN40 Dublin, Ireland; amcnabol@tcd.ie
* Correspondence: helena.ramos@tecnico.ulisboa.pt or hramos.ist@gmail.com

Received: 13 August 2020; Accepted: 18 August 2020; Published: 20 August 2020

Abstract: New challenges in water systems include different approaches from analysis of failures and risk assessment to system efficiency improvements and new innovative designs. In water distribution networks (WDNs), the risk function is a measure of its vulnerability level and security loss. Analyses of transient flows which are associated with the most dangerous operating conditions, are compulsory to grant the system liability both in water quantity, quality, and management. Specific equipment, such as air valves are used in pressurized water pipes to manage the air inside associated with the filling process, that can also act as a control mechanism, where the major limitation is its reliability. Advanced tools are developed specifically to smart water grids implementation and operation. The water system efficiency and water-energy nexus, through the implementation of suitable, pressure control and energy recovery devices, and pumped-storage hydropower solutions, provide guidelines for the determination of the most technical cost-effective result. Integrated analysis of water and energy allows more reliable, flexible, and sustainable eco-design projects, reaching better resilience systems through new concepts. The development of model simulations, based on hydraulic simulators and computational fluid dynamics (CFD), conjugating with field or experimental tests, supported by advanced smart equipment, allow the control, identification, and anticipation of complex events necessary to maintain the water system security and efficiency.

Keywords: safety and control; hydraulic transients and CFD analyses; water systems efficiency; smart water grids; water-energy nexus; energy recovery; new design solutions and eco-design

---

1. Introduction

In recent years, the lifespan of several water systems has been exceeded, so there are certain sectors from the water industry which includes drinking, storm, waste, irrigation, process industry systems, that do not reach enough operating conditions, in particular in drinking systems. This problem could produce the formation of biofilm on pipe walls [1,2], promoting bacterial growth and its transport until consumption. On the other hand, the system's own useful life, the intermittency in the supply, the leaks level, and the demands behavior cause a great variety of residence times of the drinking water and variations in the reaction of the disinfectant in the networks [3,4]. Chlorine disinfection is one of the principal factors in drinking water treatment processes, since it is used mainly to ensure the destruction of pathogen organisms that could be present in water.

Additionally, for efficient operation of water systems, suitable designs are required, which should consider economic criteria [5–8], hydraulic parameters, such as resilience [9] and water quality [10], and management criteria, such as system flexibility [11] and robustness [12]. However, a design is made based on a specific model of the water distribution network (WDN), and uncertainties in roughness [13] and mainly in future demands [14] can affect its real operating conditions. Taking into account all of these variables and uncertainties, multi-objective approaches can reveal as an alternative to reach feasible designs [15–21]. However, in water systems, cost and reliability are conflicting parameters, i.e., to improve one of them, the other has to be impaired.

Another main problem related to the operation and priming of water distribution systems is the presence of air inside the pipes [22]. There are many causes giving rise to the presence of air pockets: filling and emptying operations, temporary interruptions of water supply, vortexes in pumping feed tanks, air inlet in points with negative pressure, inflow in air valves during the negative pressure wave of a hydraulic transient and the release of the dissolved air in the water. The presence and movement of the air in water distribution pipes causes problems in most cases. Air pockets inside pipes can generate disturbances, such as the reduction of pipe cross section, even blocks, the generation of an additional head loss, which increases energy consumption of pumping groups. The decrease of pump performance is another problem, with the loss of efficiency, appearance of noise and vibration problems, corrosion inside pipelines, and significant errors in flow meters or other instrumentation equipment [23,24]. These problems may also lead to an irregular system operation and extreme surge pressure caused by entrapped air pockets.

Worldwide, water companies in all water sectors use smart equipment, such as some types of flowmeters to measure and at the same time control leakages, excess of pressure, and the amount of water consumed. This and other equipment are quite vital for smart management since improvements are made in the acquisition, storage, and treatment of data collected that also influences several water systems' performance indicators regarding water and energy balances [25,26]. The increasing need for energy in current societies is inducing more emissions of carbon dioxide to the atmosphere worsening the climate change issues. For that reason, the use of renewable energies has received excellent acceptance in recent years towards the carbon neutralization, inducing the increase of several innovative solutions [27–29]. Ensuring a clean environment and sustainable development, renewable energy sources are widely and globally appointed as future targets with great interests in hydro, wind, and solar as main green energy sources, where hydropower is considered as one of the most flexible solution for the integration of other renewables. Therefore, the idea of power production using water based on its available flow energy can contribute to the reduction of significant environmental impacts [28,29].

This Special Issue comprises papers focused on the most important issues related with new challenges of water systems, such as:

- Safety and surge or disturbances control in water systems, presenting a summary of their most recent work on research, in advanced tools, with integrated methodologies, in managing assessment, new applications, modelling, and experimental test results towards more sustainable water systems, in any type of the water sector.
- Integrated analyses evidence the new challenges in hydraulic operation and computational fluid dynamics (CFD) simulations, which are increasingly expected to develop in the near future, with interesting lessons to learn and share in the definition of operating rules, types of maneuvers, and categorization of each system response.
- Another subject is regarding the water systems' efficiency, with case studies as examples of how it needs to get better results since water and energy efficiency are interconnected and are main variables in the water sector, since, in many cases, water uses energy to be supplied and energy uses water to be produced.
- A new issue is associated with smart water grids, which are linked to the former ones too, in a holistic point of view with suggestions on how to make each system smarter and more proficient.
- The urgency of new challenges related to water-energy nexus and energy recovery in this new era for sustainable and eco-design solutions, due to restrictions imposed by climate change, calls for more flexible and new adaptive answers.

Hence, the following section summarizes the contributions according to this categorization.

2. Contributed Papers

2.1. Safety and Control

Several trends of safety research can be identified, such as risk analysis, where assessment models form the basis for building decision models, where the assessment of design options in terms of technical safety issues, is the basis for choosing the best result. The loss of safety in water systems may result directly from the failure of its individual subsystems or elements, such as water intakes, pumping stations, the water distribution network (WDN) or its utilities; from the failure of other technical systems (e.g., sewerage, energy, water structures); from undesirable extreme natural phenomena like floods and droughts; from the incidental pollution of water sources. Risk analysis of water systems should be preceded by analysis of the reliability of all subsystems in terms of interruptions to water supply, as well as failure to meet quality requirements for health posing threats to consumers of water.

Risk acceptance criteria can be used in the decision-making process regarding the operation of such a system. These criteria are related to the reliability of the system operation, in terms of both quantity and quality, in accordance with applicable rules, as well as social and economic conditions. A proposed Modified Multi-Criteria Decision Analysis Implementing an Analytic Hierarchy Process for Risk Assessment as regards failures in a WDN is presented. This procedure entails a choice of criteria influencing the risk of failure in a WDN, and the future occurrence thereof. Another important step in the procedure overall is the selection of relevant alternatives and the determination of implications in the defined criteria. Several categories and subcategories of criteria for the analysis and identification of areas at risk of WDN failure can be identified based on design, performance, operation, social impacts, financial losses, environmental, and surrounding impacts.

The water quality needs to verify the fluid interaction between disinfectant and contaminant that can occur along WDN which is dominated by convection, analyzing the variation of Turbulent Schmidt Number vs. experimental tests. More accurate mixing models improve the water quality simulations to have an appropriate control for chlorine and possible contaminants in water systems.

On the other hand, air inside a pressurized flow requires careful operational attention to understand its behavior (Figure 1), because characteristic curves of the ratio between the admitted/ejected airflow and the difference in pressure between the inside and outside of an air valve obtained with experimental tests in all possible operating regions (Figure 1) are not always known correctly.

Unfortunately, the available conditions under air valves are nearly impossible to reproduce in lab conditions with sufficient reliability. The use of CFD models can be used, not only to analyze specific hydraulic elements in water systems, but also to verify the source of flow problems, assessing the hypotheses drawn by operator experts, and to identify inconsistences regarding flow measurements in real hydraulic circuits. Intensive studies stated that several errors are mostly associated to flowmeters, and the low accuracy is connected to the perturbations induced by air and the system layout.

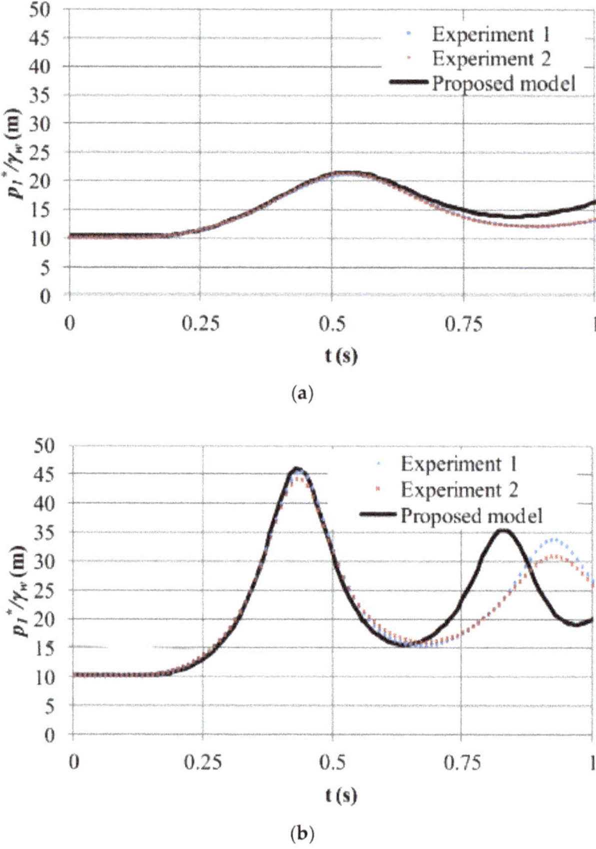

(a)

(b)

Figure 1. Air pocket pressure patterns: for initial air pocket sizes (x0) 1.36 m, and initial gauge pressures (p0) of 0.50 (**a**) and 1.25 (**b**) bar in the hydro-pneumatic tank during a filling process.

2.2. Hydraulic Transients and CFD Analyses

A measurement of the resilience is the capacity to meet demands during emergency situations when pipe bursts can occur. An emitter coefficient is added to each node to simulate a leak flow that is caused by a pipe burst percentage of the total inflow. The behavior of resilience, according to the size of the main pipe system and the node where the pipe burst occurred can be identified. In a WDN pipes that are far from the feed reservoirs have smaller diameters and their flows are very low. On the other hand, pipes close to the reservoir are well sized, and have enough capacity to support demand variation. Finally, it is noticeable that for big diameters, resilience is less affected by a pipe burst. This can be confirmed when comparing Figure 2a, where the pressure zones for the highest consumption period are shown for the initial scenario, and with big diameters trunk network of 20% and 60% of the pipes, even with leakage. The rehabilitation significantly increases the pressure in the entire network. In addition, when the pipe burst is simulated, the pressure drop is reduced but the minimum value needs to be kept. The flow behavior inside the pipe system can induce some measurement uncertainties, usually taken for granted, that can be identified using advanced analyses based on hydraulic models and CFD simulations (Figure 2b).

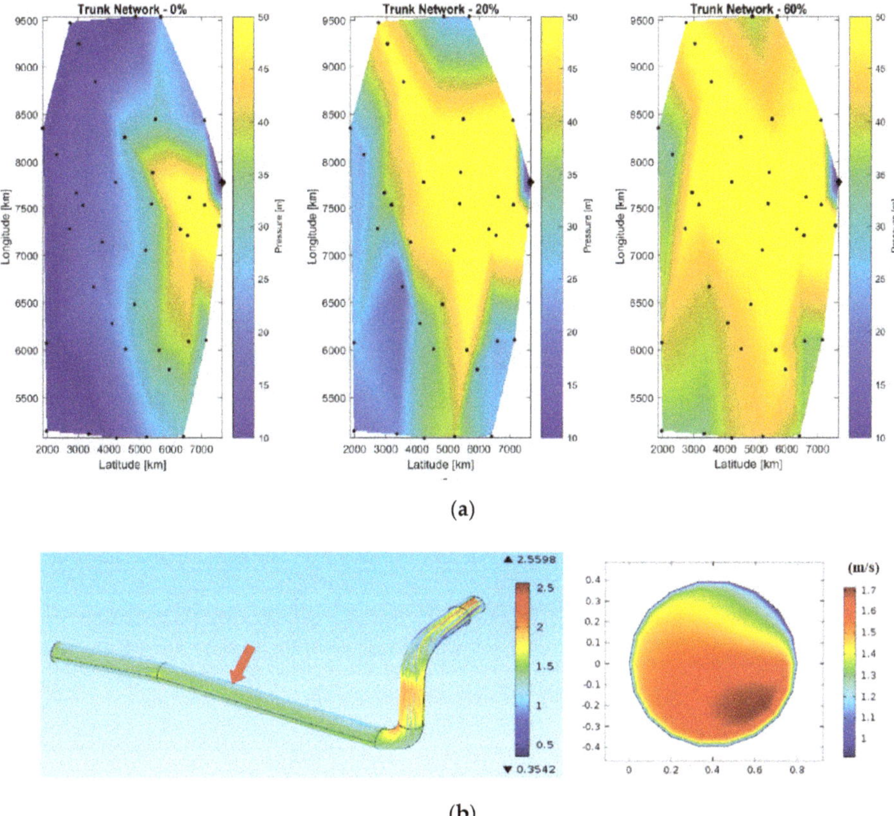

Figure 2. Pressure zones for the initial scenario and for trunk mains with 20% and 60% of the pipes with bigger diameters and a pipe burst (**a**); and streamlines and velocity distribution in the cross section of a pipe (**b**).

2.3. Water Systems Efficiency

The flow measurements can be correlated to the system efficiency. Usually, the systems are in part driven by gravity and by pressure differences, which require a pumping station. If the measurement accuracy is guaranteed (Figure 3), a higher energy efficiency level is possible to achieve, making possible a working period plan in the lower energy tariffs depending on the regularization ability and the water needs downstream.

Pressure reducing valves (PRVs) are a convenient device for reducing leakage by pressure control. However, the energy dissipation that takes place in a PRV is wasteful of energy resources, and this energy could be recovered by substituting or putting in series, the PRV with a micro-turbine. Thus, in addition to the reduction of water losses, a certain part of the energy in the network could be recovered, reducing greenhouse gas emissions and making the water supply system more sustainable [14]. Due to the potential of hydropower in these systems, several investigations have focused on the evaluation of the installation of these devices in water networks, showing that up to 40% of the gross power potential available in a PRV could be recovered by replacing/coupling the PRV by/with a PAT, depending on the manager objectives.

Some case studies presented show results achieved about the implementation of measures for the monitoring and water loss control, which allowed accessing a higher level of efficiency, especially in the reduction of water losses, in saving energy, and the consequent reduction of associated costs.

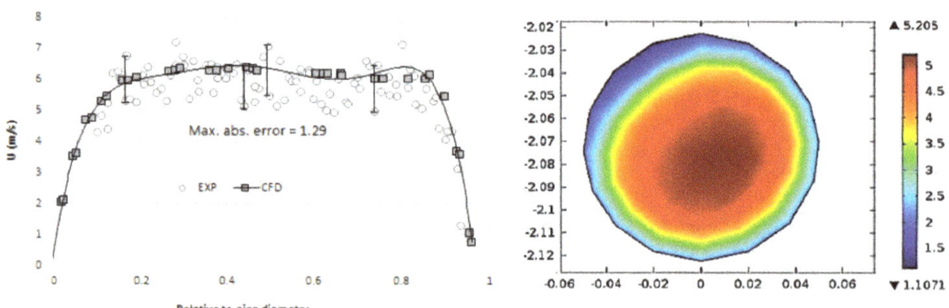

Figure 3. Flow velocity distribution and implications on the system balance efficiency.

2.4. Smart Water Grids

Water management towards smart grids and cities is an issue increasingly appreciated under financial and environmental sustainability focus in any water sector, disclosing the technological breakthroughs associated with water and energy use. The water industry is subject to new challenges regarding the sustainable management of urban water systems (Figure 4). There are many external factors, including impacts of climate change, drought, and population growth in urban centers, which lead to an increase of the responsibility, in order to adopt more sustainable management of the water sector. Smart water management aims at the exploitation of water, at the regional or city level, on the basis of sustainability and self-sufficiency. This exploitation is carried out through the use of innovative technologies, such as information, control technologies, and monitoring.

The development of smart techniques requires technology use in water systems, as well as its implementations. They will improve the performance of many networks characterized by degraded infrastructure, irregular supplies, and low levels of customer satisfaction or substantial deviations of the proportional bills to real consumption. A smart water system can lead to more sustainable water services, allowing to reduce financial losses, enabling innovative business models to serve better the urban and rural populations.

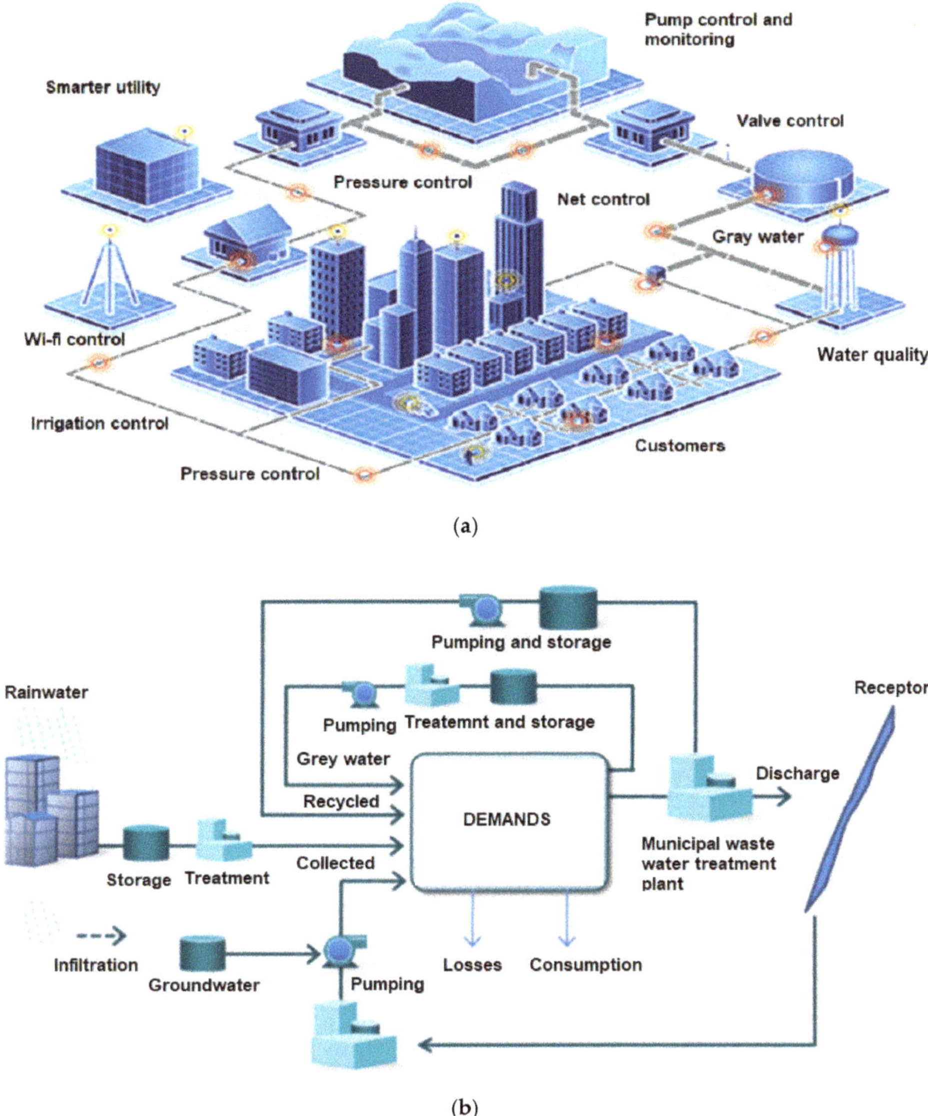

(a)

(b)

Figure 4. Holistic view of smart grids: integration of components in a smart water grid in urban environment (**a**); and water cycle process in a smart water grid (**b**).

2.5. Water-Energy Nexus

Water abstraction and distribution are among the activities in which the water-energy nexus plays an important role. In the European Union, as an example, 8% of the total energy consumption, in recent years, was related to the water supply. In addition, it is estimated that 32 billion cubic meters per year (66% of the treated water) are lost in the water distribution process globally. This is mainly due to the ageing infrastructure, the non-optimal design of the water supply systems, and the increase in water stress, in urban areas. Different strategies have been proposed to reduce energy consumption from fossil fuels in the water industry sector, by using renewable energy sources, available flow energy, or by recovering the excess of heat at the wastewater treatment plant. Leakage management can also play a major role in the reduction of energy consumption in the water sector. Thus, pressure control in water distribution networks is one of the most effective measures to reduce leakages, because of the direct relation between pressure and leakage rate. Hence, Pressure Reducing Valves (PRVs) can reduce water losses by pressure control. The energy dissipation that takes place in a PRV is wasteful of energy resources, and this energy could be recovered by substituting the PRV with a micro-hydro device. Thus, in addition to the reduction of water losses, a certain part of the energy in the network could be recovered, reducing greenhouse gas emissions, the water footprint, and making the water supply systems more sustainable. Due to the potential of hydropower in these systems, several investigations have focused on the evaluation of the installation of these devices in water networks, showing that up to around half of the gross power potential available in a PRV could be recovered by replacing the PRV with a PAT (Figure 5). Other potential locations for installing hydropower turbines are in break pressure tanks, water storages, or water treatment plants, allowing the estimation of significant energy generation potential in some locations of water systems.

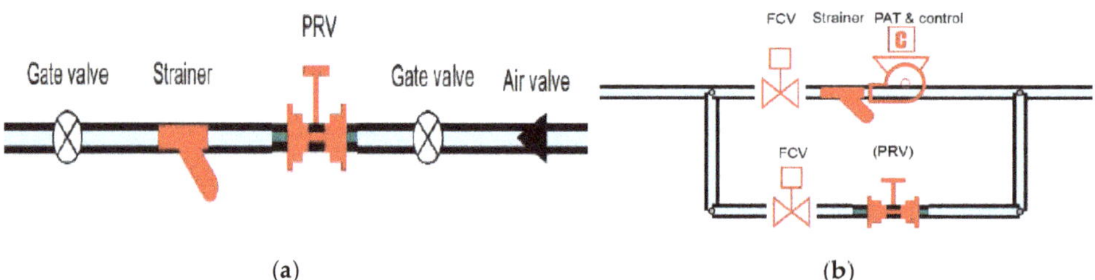

Figure 5. Installation scheme of a pressure reducing valve (PRV) (**a**) and a pump as turbine (PAT) (**b**).

2.6. Energy Recovery

A particular class of micro-turbines consists of pumps working in reverse mode, i.e., pump as turbines (PAT). These are devices that can be installed along distribution pipes to reduce pressure at nodes and recover energy, with significantly reduced investment costs compared to traditional turbines. PATs are considered to be a cheaper technology compared to traditional turbines for small hydropower energy recovery. However, information related to the total PAT cost, also including installation cost, is not easily accessible. Overall, methodologies focused on the use of PATs took into account a PAT cost according to the generated power. For some analyzed cases, the replacement of PRVs with PATs would involve powers between 400 W and 2 kW and when the PAT cost was considered between 50% and 10% of the total costs, showed a payback period between 3 and 6 years, respectively (Figure 6).

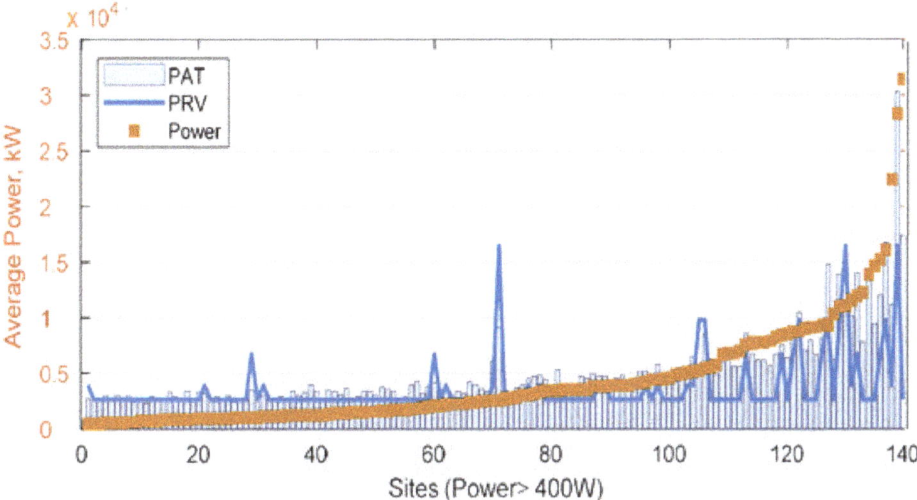

Figure 6. Micro-hydro average generated power by PATs (red) and comparison between PAT vs. PRV total cost (blue).

2.7. New Design Solutions and Eco-Design

Energy and climate change are thoroughly linked, since fossil energy generation highly affects the environment, and climate change influences the renewable energy generation capacity. There are studies that give a new contribution to the energy generation in water infrastructures by means of new concept installations, in particular an inline pumped-storage hydro (IPSH) solution. Increasingly, there are great interests in wind and solar as green energy sources, and hydropower is seen as a huge flexibility. Currently, hydropower is considered as one of the most preferred sources to produce electricity and, simultaneously, for integration of other renewable sources.

It is worth mentioning that the pump consumes energy and the hydropower produces it creating a new loop system adapted to existing infrastructures with direct flow condition, based on the available head, a by-pass line, that can be activated to use the head difference for energy generation.

Therefore, the idea of power production using water based on its available flow energy can contribute to the reduction in significant environmental impacts (Figure 7). The application of MHP solutions has gone even further to different water sectors, e.g., irrigation networks with a promising future of their applicability. Additionally, novel solutions using the compressibility effect of air have been presented in some studies that can be combined with pumped-storage hydropower (PSH and ACUR) to offer a hybrid solution.

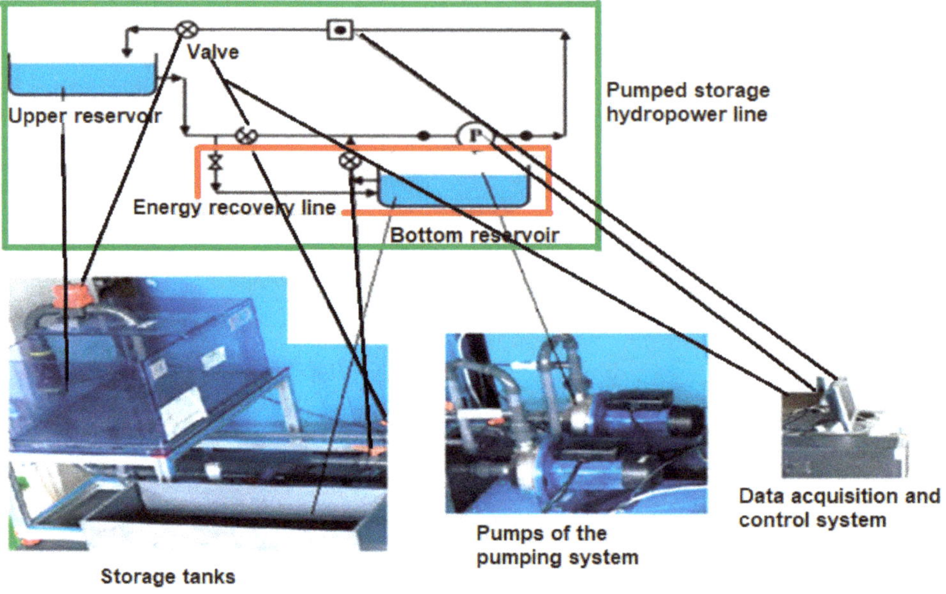

(**a**)

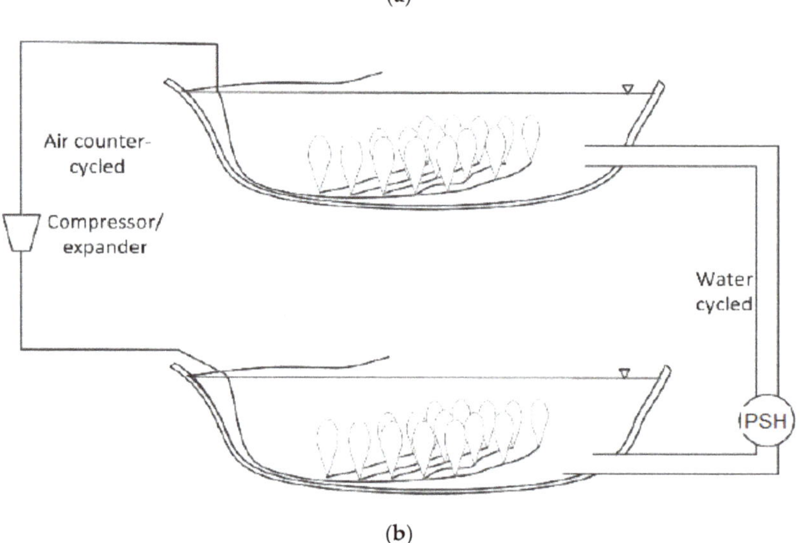

(**b**)

Figure 7. Pumped-storage hydropower integration in water systems (**a**) and with an Air Cushion Underground Cavern (ACUR) (**b**).

3. Conclusions

Risk analysis is found to contribute significantly to the reduction of failure overall, identifying and limiting their causes and effects. If numerous overall failures arise in a water distribution network (WDN), the water company has to modernize and renovate the system, so as to minimize leakages, bursts, and other pipeline failures. A database of weak points of system functioning, with particular emphasis placed on the frequency of occurrences and associated negative effects, need to be registered. In safety management, the selection of risk protection measures and its implementation into operational practice needs to be carried out for the effectiveness of applied solutions. Residence Time Distribution (RTD) curves are good features and a novel proposal on mixing cross-junctions, in order to calibrate simulation processes of water quality analysis, with the help of stimulus-response techniques. Finding solutions to the problem of inadequate use of chlorine is an issue that involves various aspects. When WDN presents a high degree of reaction due to pipe conditions and residence times, concentrations in some areas decrease rapidly and can affect costumer health.

In addition, a system that is capable of operating only with minimum pressure values is not sufficiently robust to surpass emergency situations, such as pipe bursts or firefighting, causing water shortage in some areas. To avoid this problem, the resilience is an important parameter to be considered, since it reflects the network capacity to cope with a variety of events. The methodology explicitly quantifies the benefit of resilience through an energy recovery solution. During normal operation, pumps as turbines (PATs) operate to perform pressure control, while, in emergency scenarios, the flow can be driven through a by-pass to regulate pressure at critical points. For existing networks, the rehabilitation only of the main pipes is an alternative to improve resilience, since these pipes are the most relevant for the system.

CFD techniques provide a tool for determining in detail the behavioral characteristics of hydraulic elements. However, CFD requires a proper calibration of the models, so it will be essential to validate the results by tests in field or laboratory. Specifically, the effectiveness of these techniques to represent the behavior of air valves, bends, cross sections, pumps, turbines, and other hydraulic elements that are installed in water systems are analyzed.

In some cases, the measurement problem has a significant influence on the correct management of technical, hydraulic, and economic efficiency of water companies. Since the hydraulic circuits have, normally, flowmeters, a water balance is necessary. These balances are important tools to detect leaks throughout the supply and distribution processes by assessing that the installation requirements proposed by manufactures are not sufficient to dissipate such detected uncertainties.

Regarding the excess of pressure in water systems, the replacement or conjunction of PRVs with PATs as a measure to generate energy and reduce leakage represents an efficient strategy to improve the sustainability of water systems. This is a challenge to control pressure and generate energy at the same time, also analyzing the most convenient element from technical and economic points of view, highlighting the potential of micro hydro power (MHP) installations in general and PATs in particular, as a tool to reduce electricity costs and improve the sustainability of water systems.

Some of the main advantages of smart water management are a better understanding of the water system, namely the detection of leaks, conservation and monitoring of water quality. The implementation of smart water system technologies enables public service companies to build a complete database for the identification of the areas where water losses or illegal connections occur. The advantages of smart water grids are economic benefits to water and energy conservation, while the efficiency of the system can improve customers service. Some of the main smart technologies are listed as follows: (i) smart pipe and sensor; (ii) smart water metering; (iii) Geographic Information System (GIS); (iv) cloud computing and supervisory control and data acquisition; (v) models, tools of optimization, and decision support systems.

The contributors to this Special Issue are invited to provide a series of approaches to create innovation, practical implementation, and further awareness and solutions by presenting research based on case studies with integrated approaches, advanced tools, and water management practices, which allow to enhance:

1. Innovative water network analyses, safety solutions, and available tools, through specific studies in new or existing water infrastructures of small to big scales. This does not require significant changes and bulky investments to increase their efficiency, flexibility, resilience, robustness, and feasibility;

2. The energy generation using the gravitational flow appears as a significant decentralized economic advantage in the definition of the energy recovery solutions;

3. Depending on the type of demand, the applications show smart pressure and flow control measures for new challenges in future water systems including water-energy efficiency solutions;

4. The smart approach based on a controlled water and consequently energy recovery solution, increases water systems flexibility, contributing to smart management and water and energy efficiency in the water sector. These procedures allow to better face the existing costs associated with pumping, treatment plants, water leakage, expansion and reparation of infrastructures, and water bills for customers.

Author Contributions: H.M.R. conceived and led the development of this Special Issue and this paper; A.C. and A.M.N. contributed substantially to the writing. All authors have read and agreed to the published version of the manuscript.

Funding: This research received no external funding.

Acknowledgments: The authors of this paper, who served as guest editors of this Special Issue, wish to thank the journal editors, all authors submitting papers to this Special Issue, and the many referees who contributed to paper revision and improvement of all published papers.

Conflicts of Interest: The authors declare no conflict of interest.

Acronyms

ACUR	Air Cushion Underground Cavern
CFD	computational fluid dynamics
GIS	geographic Information System
IPSH	inline pumped-storage hydro
MHP	Micro-hydropower
PAT	Pump as turbine
PRV	Pressure reducing valve
PSH	Pumped-storage hydropower
WDN	Water distribution network
WSS	Water supply system

References

1. Geem, Z.W.; Kim, J.H.; Loganathan, G. A New Heuristic Optimization Algorithm: Harmony Search. *Simulation* **2001**, *76*, 60–68. [CrossRef]

2. Maier, H.R.; Simpson, A.R.; Zecchin, A.C.; Foong, W.K.; Phang, K.Y.; Seah, H.Y.; Tan, C.L. Ant Colony Optimization for Design of Water Distribution Systems. *J. Water Resour. Plan. Manag.* **2003**, *129*, 200–209. [CrossRef]

3. Suribabu, C.R.; Neelakantan, T.R.; Renganathan, N.T. Design of water distribution networks using particle swarm optimization. *Urban. Water J.* **2006**, *3*, 111–120. [CrossRef]

4. Baños, R.; Reca, J.; Martínez, J.; Gil, C.; Márquez, A.L.; Gil, C. Resilience Indexes for Water Distribution Network Design: A Performance Analysis Under Demand Uncertainty. *Water Resour. Manag.* **2011**, *25*, 2351–2366. [CrossRef]

5. Shokoohi, M.; Tabesh, M.; Nazif, S.; Dini, M. Water Quality Based Multi-objective Optimal Design of Water Distribution Systems. *Water Resour. Manag.* **2016**, *31*, 93–108. [CrossRef]

6. Marques, J.; Cunha, M.D.C.; Savić, D. Using Real Options in the Optimal Design of Water Distribution Networks. *J. Water Resour. Plan. Manag.* **2015**, *141*, 04014052. [CrossRef]

7. Schwartz, R.; Housh, M.; Ostfeld, A. Least-Cost Robust Design Optimization of Water Distribution Systems under Multiple Loading. *J. Water Resour. Plan. Manag.* **2016**, *142*, 04016031. [CrossRef]

8. Giustolisi, O.; Laucelli, D.; Colombo, A.F. Deterministic versus Stochastic Design of Water Distribution Networks. *J. Water Resour. Plan. Manag.* **2009**, *135*, 117–127. [CrossRef]

9. Lansey, K.E.; Duan, N.; Mays, L.W.; Tung, Y. Water Distribution System Design Under Uncertainties. *J. Water Resour. Plan. Manag.* **1989**, *115*, 630–645. [CrossRef]

10. Zheng, F.; Simpson, A.; Zecchin, A.C. Improving the efficiency of multi-objective evolutionary algorithms through decomposition: An application to water distribution network design. *Environ. Model. Softw.* **2015**, *69*, 240–252. [CrossRef]

11. Geem, Z.W. Multiobjective Optimization of Water Distribution Networks Using Fuzzy Theory and Harmony Search. *Water* **2015**, *7*, 3613–3625. [CrossRef]

12. Prasad, T.D.; Park, N.-S. Multiobjective Genetic Algorithms for Design of Water Distribution Networks. *J. Water Resour. Plan. Manag.* **2004**, *130*, 73–82. [CrossRef]

13. Ramos, H.; Borgå, Å. Pumps as turbines: An unconventional solution to energy production. *Urban. Water* **1999**, *1*, 261–263. [CrossRef]

14. Carravetta, A.; Houreh, S.D.; Ramos, H.M. *Pumps as Turbines: Fundamentals and Applications*, 1st ed.; Springer: Cham, Switzerland, 2017.

15. Pérez-Sánchez, M.; Sánchez-Romero, F.-J.; Ramos, H.; López-Jiménez, P.A. Energy Recovery in Existing Water Networks: Towards Greater Sustainability. *Water* **2017**, *9*, 97. [CrossRef]

16. De Marchis, M.; Freni, G. Pump as turbine implementation in a dynamic numerical model: Cost analysis for energy recovery in water distribution network. *J. Hydroinformatics* **2015**, *17*, 347–360. [CrossRef]

17. Carravetta, A.; Del Giudice, G.; Fecarotta, O.; Ramos, H. PAT Design Strategy for Energy Recovery in Water Distribution Networks by Electrical Regulation. *Energies* **2013**, *6*, 411–424. [CrossRef]

18. Lima, G.M.; Luvizotto, J.E.; Brentan, B.M.; Ramos, H. Leakage Control and Energy Recovery Using Variable Speed Pumps as Turbines. *J. Water Resour. Plan. Manag.* **2018**, *144*, 04017077. [CrossRef]

19. Carravetta, A.; Del Giudice, G.; Fecarotta, O.; Ramos, H. Energy Production in Water Distribution Networks: A PAT Design Strategy. *Water Resour. Manag.* **2012**, *26*, 3947–3959. [CrossRef]

20. Lima, G.M.; Brentan, B.M.; Izquierdo, J.; Ramos, H.; Luvizotto, J.E. Trunk Network Rehabilitation for Resilience Improvement and Energy Recovery in Water Distribution Networks. *Water* **2018**, *10*, 693. [CrossRef]

21. Pietrucha-Urbanik, K.; Tchórzewska-Cieślak, B. Approaches to Failure Risk Analysis of the Water Distribution Network with Regard to the Safety of Consumers. *Water* **2018**, *10*, 1679. [CrossRef]

22. García-Todolí, S.; Iglesias-Rey, P.L.; Mora-Meliá, D.; Martínez-Solano, F.J.; Fuertes-Miquel, V.S. Computational Determination of Air Valves Capacity Using CFD Techniques. *Water* **2018**, *10*, 1433. [CrossRef]

23. Coronado-Hernández, Ó.E.; Besharat, M.; Fuertes-Miquel, V.S.; Ramos, H. Effect of a Commercial Air Valve on the Rapid Filling of a Single Pipeline: A Numerical and Experimental Analysis. *Water* **2019**, *11*, 1814. [CrossRef]

24. Cervantes, D.H.; Delgado-Galván, X.; Nava, J.L.; López-Jiménez, P.A.; Rosales, M.; Mora-Rodriguez, J. Validation of a Computational Fluid Dynamics Model for a Novel Residence Time Distribution Analysis in Mixing at Cross-Junctions. *Water* **2018**, *10*, 733. [CrossRef]

25. Ramos, H.; McNabola, A.; López-Jiménez, P.A.; Pérez-Sánchez, M. Smart Water Management towards Future Water Sustainable Networks. *Water* **2019**, *12*, 58. [CrossRef]

26. Simão, M.; Besharat, M.; Carravetta, A.; Ramos, H. Flow Velocity Distribution Towards Flowmeter Accuracy: CFD, UDV, and Field Tests. *Water* **2018**, *10*, 1807. [CrossRef]

27. García, I.F.; Novara, D.; McNabola, A. A Model for Selecting the Most Cost-Effective Pressure Control Device for More Sustainable Water Supply Networks. *Water* **2019**, *11*, 1297. [CrossRef]

28. Storli, P.-T.; Lundström, T.S. A New Technical Concept for Water Management and Possible Uses in Future Water Systems. *Water* **2019**, *11*, 2528. [CrossRef]

29. Ramos, H.; Dadfar, A.; Besharat, M.; Adeyeye, K. Inline Pumped Storage Hydropower towards Smart and Flexible Energy Recovery in Water Networks. *Water* **2020**, *12*, 2224. [CrossRef]

© 2020 by the authors. Licensee MDPI, Basel, Switzerland. This article is an open access article distributed under the terms and conditions of the Creative Commons Attribution (CC BY) license (http://creativecommons.org/licenses/by/4.0/).

 *water*

Article

Trunk Network Rehabilitation for Resilience Improvement and Energy Recovery in Water Distribution Networks

Gustavo Meirelles [1,*], Bruno Brentan [2], Joaquín Izquierdo [3], Helena Ramos [4] and Edevar Luvizotto, Jr. [1]

[1] Department of Water Resources—DRH, Universidade Estadual de Campinas, Campinas 13083-889, Brazil; edevar@fec.unicamp.br

[2] Centre de Recherche en Controle et Automatique de Nancy—CRAN, Université de Lorraine, 54000 Nancy, France; brunocivil08@gmail.com

[3] Institute for Multidisciplinary Mathematics, Universitat Politècnica de València, Camino de Vera s/n Edif. 5C, 46022 Valencia, Spain; jizquier@upv.es

[4] Civil Engineering, Architecture and Georesources Departament—CERIS, Instituto Superior Técnico, Universidade de Lisboa, 1049-001 Lisboa, Portugal; hramos.ist@gmail.com

*   Correspondence: limameirelles@gmail.com; Tel.: +55-(19)-98456-3199

Received: 28 April 2018; Accepted: 23 May 2018; Published: 25 May 2018

Abstract: Water distribution networks (WDNs) are designed to meet water demand with minimum implementation costs. However, this approach leads to poor long-term results, since system resilience is also minimal, and this requires the rehabilitation of the network if the network is expanded or the demand increases. In addition, in emergency situations, such as pipe bursts, large areas will suffer water shortage. However, the use of resilience as a criterion for WDN design is a difficult task, since its economic value is subjective. Thus, in this paper, it is proposed that trunk networks (TNs) are rehabilitated when considering the generation of electrical energy using pumps as turbines (PATs) to compensate for an increase of resilience derived from increasing pipe diameters. During normal operation, these micro-hydros will control pressure and produce electricity. When an emergency occurs, a by-pass can be used to increase network pressure. The results that were obtained for two hypothetical networks show that a small increase in TN pipe diameters is sufficient to significantly improve the resilience of the WDN. In addition, the value of the energy produced surpasses the investment that is made during rehabilitation.

Keywords: trunk network; water distribution network; resilience; optimization; energy recovery; pumps as turbines

---

1. Introduction

For efficient operation of Water Distribution Networks (WDNs), suitable designs are necessary, which should consider economic criteria [1–3], hydraulic parameters, such as resilience [4] and water quality [5], and management criteria, such as system flexibility [6] and robustness [7]. However, a design is made based on a specific model of the WDN, and uncertainties in roughness [8] and mainly in future demands [9] can affect its real operation conditions. Taking into account all of these variables and uncertainties, multi-objective approaches reveal as an alternative to reach feasible designs [10–12]. However, WDN cost and reliability are conflicting parameters, i.e., to improve one of them, the other has to be impaired. The importance of resilience to achieve a reliable system is a well-known fact. However, it is hard to economically quantify the value of its improvement. As a result, even in a multi-objective design, it is up to the decision maker to choose the best possible scenario based on his or her experience.

An alternative to better seize the resilience improvement is the use of Pumps as Turbines (PATs) to generate electrical energy during the periods of normal operation conditions, when those micro-hydros act as both pressure control and energy generation devices simultaneously [13–15]. When an emergency occurs, such as a pipe burst, the pressure in the network can be controlled while using a by-pass. In this scenario, the additional investment that is necessary to improve resilience is recovered during normal operation conditions through energy production. Due to the dynamic behavior of a WDN, PATs are not able to perform pressure control as Pressure Reducing Valves (PRVs) [16]. To solve this problem, variable speed drives [17,18] and a combined configuration of PATs and PRVs [19,20] can be used.

The studies presented in [21,22] show the feasibility of increasing network pipes diameters to improve energy production for various topologies. However, these studies are restricted to the design of new WDNs. For existing systems, the costs of replacing pipes are much higher, since pipes are already buried. In addition, technical issues, such as traffic changes and the interruption in water supply can cause troubles for the population. In this case, rehabilitation of the Trunk Network (TN) can be an alternative to improve WDN flow capacity. According to [23,24], WDN sensitivity is strongly related to TN performance due to the high flow transported, highly influencing pressure in the entire network.

Therefore, when considering the difficult to evaluate resilience in WDN design and the high investment that is necessary to improve this parameter in existing systems, in this research, TN rehabilitation is proposed to improve WDN resilience while considering the use of PATs for energy recovery during normal operation to compensate the investment that is made. In this way, it is expected that a technical issue such as resilience can be better exploited as a design parameter, and benefits will derive for the economic side. Thus, the decision maker, even inexperienced in this kind of problem, may have a reference to choose the best alternative. Different sizes of TNs are evaluated to access the solution with minimum interventions and reasonable resilience improvement. The TN is defined using graph theory to achieve the shortest path of flow to supply a given demand node. The flow that is observed in the pipes is used as a weighting coefficient to avoid the inclusion of small pipes in TNs. Then, applying the Particle Swarm Optimization (PSO) technique to obtain the diameters of the TN pipes, allows for minimizing the implantation costs, which are calculated as the sum of pipe and micro-hydro costs, discounting the energy recovery benefits that were obtained during the PAT life span. The proposed methodology is applied in two different case studies: the Fossolo Drinking Network [25] and the Balerma Irrigation Network [26].

2. Materials and Methods

2.1. Trunk Network

A WDN topology can be seen as a graph, with the nodes representing the vertices and the pipes the edges. Several studies show that the algorithms and concepts that were developed within graph theory can be applied to define the best path of a TN [27,28].

The concept of shortest path can be used for this purpose, since it defines the frequency of all the possible paths in a graph. The flow direction in each pipe is necessary to create a so-called digraph, which is a graph with directions that are previously set. However, in a WDN, the flow behavior can change according to the day time and the settings of valves and pumps. Usually, a hydraulic simulation is made for the maximum consumption period, with the usual setting of pumps and valves to define the flow direction. Taking into account tanks and reservoirs as sources for water supply, the Breadth First Search (BFS) technique [29] can be applied to define the shortest path in terms of nodes between the source and each consumption node. Then, a square matrix is built, with columns representing the start nodes and rows the end nodes. The sum of the values of each row, which are called Accumulated Shortest Path Value (ASPV), will define the most important nodes of the WDN, since this sum indicates their reach, i.e., to meet the demand of some areas the flow has to pass through these most important nodes. The values have to be first normalized to achieve this importance classification.

According to [23], an additional criterion is necessary to avoid the introduction of small pipes that are capable of supplying larger pipes in the TN. Thus, the flow that was observed in the pipe is used as a weighting coefficient for the ASPV, and only really important pipes will present high ASPVs.

Finally, the size of the TN has to be defined. A statistical analysis of the frequencies of ASPVs has to be done to achieve the best value. However, in this work the optimal size of a TN is defined according to the value of an objective function. Thus, to evaluate different scenarios, the TN size is gradually changed, and the results are compared in order to define the best configuration. Figure 1 summarizes the procedure to obtain the TN.

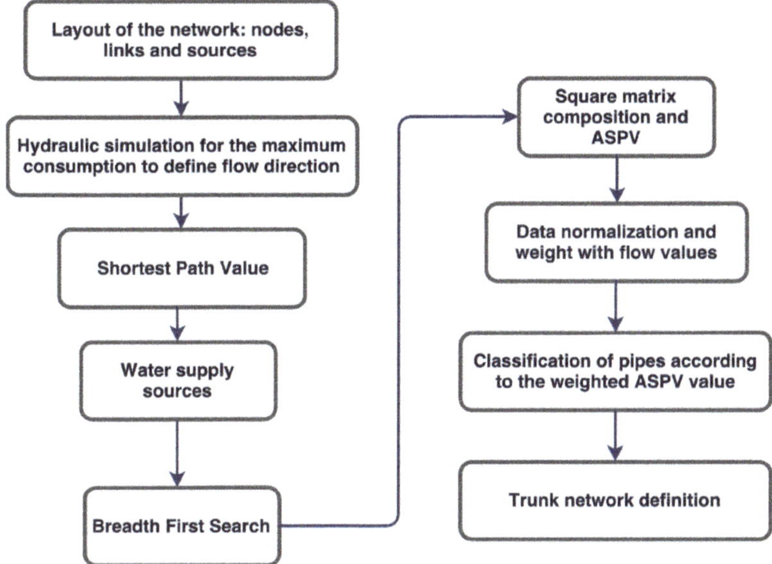

Figure 1. Flowchart to obtain the Trunk Network.

2.2. Resilience Index

The resilience index that is proposed by [30] expresses the capacity of a WDN to supply the demand in emergency situations, such as pipe bursts, pumping station shutdowns, and even sudden demand increases, as observed in firefighting. In all of these cases, the velocity in certain pipes increases, and the additional headloss produced causes pressure reduction in the consumption nodes. Thus, if some additional pressure is not available, then the minimum conditions for supply are not met, causing water shortage in some network zones. In the proposed scheme, the WDN operates close to these minimum conditions during a normal scenario. However, in emergency cases, the flow can be driven through a by-pass to regulate the pressure in the network, thus reducing the head drop that is caused by the PAT. Two valves are necessary to isolate and regulate the PAT flow and one valve is used in the by-pass, as shown in Figure 2. With this configuration, the flow can be regulated in both branches, and in extreme conditions, only through the by-pass. In addition, the PAT isolation valves allow for its maintenance when necessary. The resilience index will be calculated for the condition where the PAT is not operating, according to Equation (1).

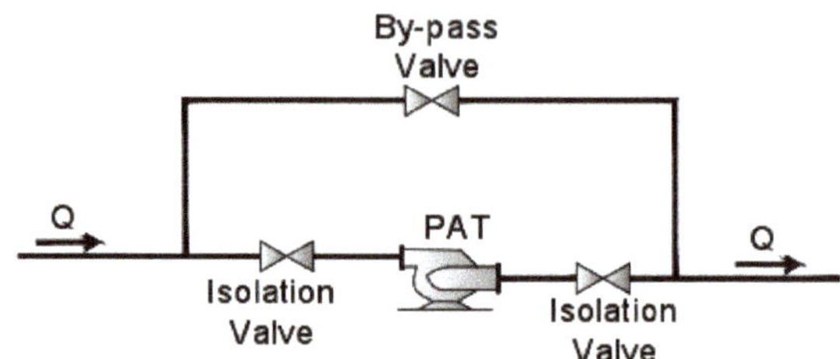

Figure 2. Hydraulic layout of Pumps as Turbines (PAT) and by-pass.

$$I_r = \frac{\sum_{i=1}^{N_n} q_i \cdot \left(h_i - h_{req_i}\right)}{\left(\sum_{k=1}^{N_r} Q_k \cdot H_k + \sum_{j=1}^{N_p} Q_j \cdot H_j\right) - \sum_{i=1}^{N_n} q_i \cdot h_{req_i}} \tag{1}$$

Here I_r is the resilience index, N_n is the number of demand nodes, q_i is the demand of node i, h_i is the hydraulic grade available in node i, h_{req_i} is the hydraulic grade required by node i, N_r is the number of reservoirs and tanks, Q_k is the flow supplied by the reservoir or tank k, H_k is the hydraulic grade of reservoir or tank k, N_p is the number of pumps, Q_j is the flow that is supplied by pump j, and H_j is the head of pump j. As the most unfavorable scenario occurs during maximum consumption, resilience has been calculated only for this condition.

2.3. Trunk Network Design and Energy Recovery

The design of the TN can be made through the minimization of the objective function, OF, defined in Equation (2), which expresses the net cost of the project. The total cost, which is represented by the pipes' cost, CP, and the micro-hydro cost, CM, is discounted by the benefit that is obtained from the energy recovery during the PAT life span, CE.

$$OF = \min(CP + CM - CE) \tag{2}$$

Both costs (of pipes and micro-hydros), which include design, civil works, and electro-mechanical equipment, were obtained from Brazilian companies, for different sizes, as shown in Figure 3. Equations (3) and (4) can be used to calculate pipes and micro-hydro costs, which are valid for 25 to 600 mm and 5 to 224 kW, respectively. Figure 4 shows the dimensionless curves of PATs, where the head, H, and flow, Q, conditions are related to the operation on its Best Efficiency Point (BEP). A set of 14 curves was available inside a specific speed range of 0.46 to 4.94.

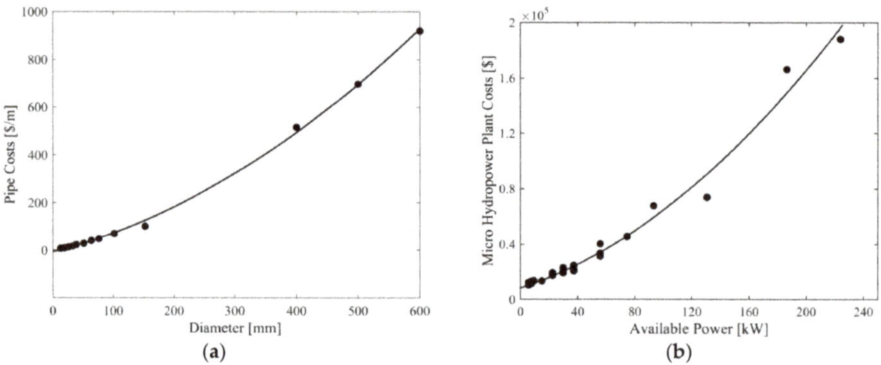

Figure 3. Design costs: (**a**) Pipes; and (**b**) Micro hydro [22].

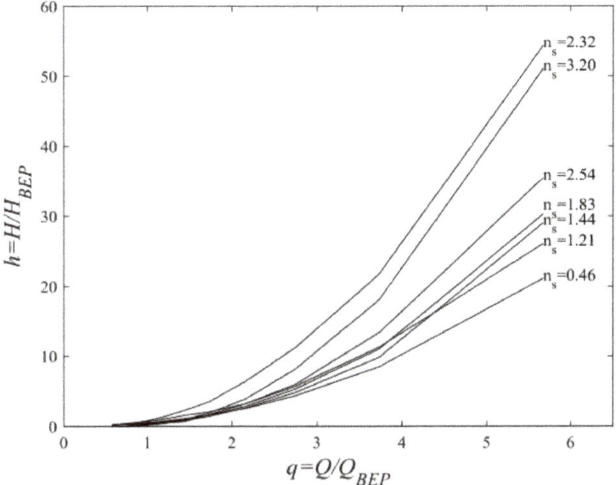

Figure 4. Dimensionless curves of PATs [18].

$$CP = \sum_{i=1}^{NT} \left(0.001539 \cdot DC_i^2 + 0.635491 \cdot DC_i - 7.107225\right) \cdot L_i \tag{3}$$

$$CM = 2.246 \cdot P^2 + 338.23 \cdot P + 8218.79 \tag{4}$$

Here, NT is the number of pipes of the TN, DC is the unitary cost of pipe i, L the length of pipe i, and P is the power of the micro hydro.

The benefit that was obtained from the energy recovery during the PAT life span is calculated using the Net Present Value (NPV), as defined in Equation (5).

$$CE = NPV = \sum_{t=1}^{n} \frac{\left(\sum_{i=1}^{8760} (\rho \cdot g \cdot Q_{i,t} \cdot H_{i,t} \cdot \eta_{i,t}) \cdot TE_{i,t}\right) - CO_t}{(1+j)^t} \tag{5}$$

Here, n is the PAT life span, ρ is the water specific mass, g is the gravitational acceleration, $Q_{i,t}$ is the PAT flow during hour i of year t, $H_{i,t}$ is the PAT head during hour i of year t, $\eta_{i,t}$ the PAT efficiency during hour i of year t, $TE_{i,t}$ the energy tariff during hour i of year t, CO_t operation and maintenance costs of year t, and j is the annual interest rate.

2.4. Optimization and TN Design Procedure

To solve the optimization problem, the PSO technique is used. Developed by [31], it is widely used to solve a variety of problems of water supply systems, such as sectorization [32], PAT selection and location in WDNs [33], and water demand forecasting [34]. Based on group behavior, the technique searches feasible solutions while considering individual and group information. The position X of each particle represents a solution. Each particle has a velocity V that defines its search direction, and it is calculated through Equation (6), where its best position ever found, P, the best position found by the group, G, and an inertia factor, ω, are considered. Two coefficients, c_1 and c_2, respectively, defined as cognitive and social coefficients, are dynamically adjusted through iterations to allow for an extensive search in the beginning (higher value of c_1), and a refined local search at the end (higher value of c_2). At the end of each iteration, the particle position is updated using Equation (7).

$$V_i^{k+1} = \omega \cdot V_i^k + c_1 \cdot rand_1 \cdot \frac{\left(P_i^k - X_i^k\right)}{\Delta t} + c_2 \cdot rand_2 \cdot \frac{\left(G - X_i^k\right)}{\Delta t} \tag{6}$$

$$X_i^{k+1} = X_i^k + V_i^{k+1} \cdot \Delta t \tag{7}$$

The TN design will follow a three-stage procedure. First, the optimal design of the WDN, when considering the classical approach of minimizing pipes diameters (cost) that guarantee a minimum pressure, will be done. This design will be used as the default operation condition of the network. Then, the TN will be defined using the ASPV weighted by pipes flows. The size of the TN will be defined in a range of 10–100% of the total number of pipes of the network in order to verify the improvement of resilience and energy production with the TN diameter increase. In the last stage, for each scenario, the TN will be designed again as a rehabilitation process, with a pipe cost that is 20% higher to account for replacement difficulty. In this second design stage, in addition to pipe diameters, PAT operation points (head and flow) are also variables to be defined. Variable speed operation is considered with a speed constraint of 60–100% of its nominal value, and a flow constraint up to 50% of its nominal value [18]. As a result, it is expected that the best configuration of pipes and PATs will be found, maximizing the energy production. In all cases, the software EPANET [35] was used to model the network and to simulate the various scenarios to achieve flow and pressure data. Figure 5 summarizes the described TN design procedure, including PAT selection.

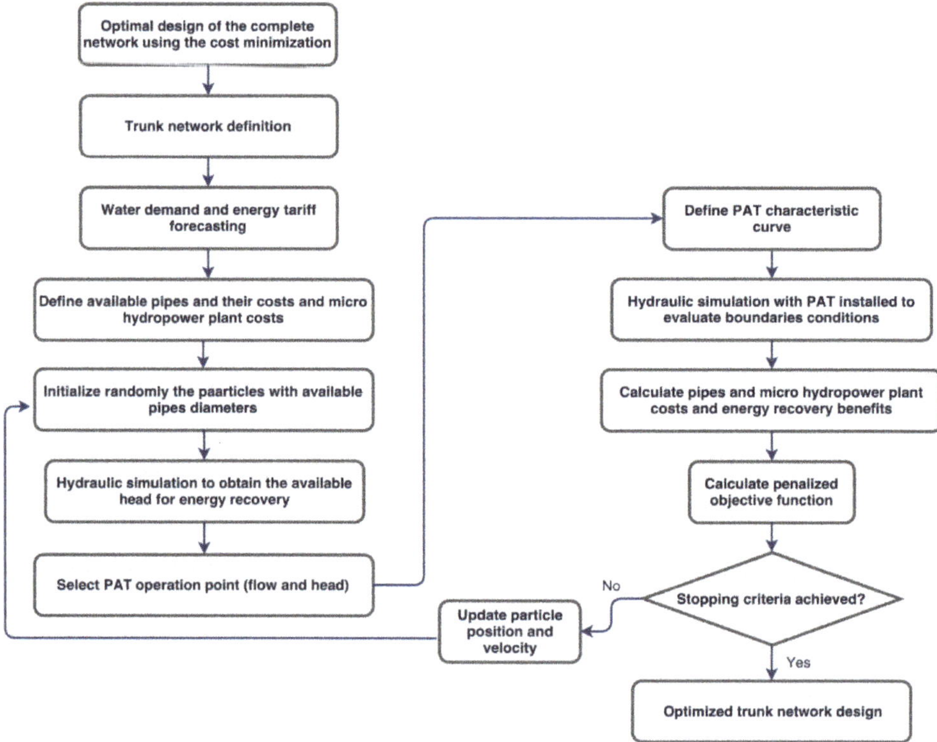

Figure 5. Flowchart of Trunk Network (TN) design with PAT selection.

3. Results

3.1. Fossolo Network

The first case study considers the Fossolo drinking network [25]. It has 58 pipes and 36 consumption nodes with one reservoir for water supply, where a PAT is installed, as shown in Figure 6. A minimum pressure of 10 m is considered for the design, as set by the Brazilian standards [36].

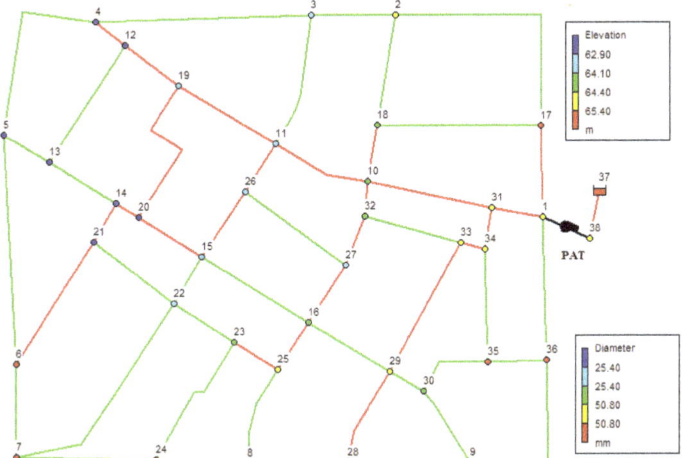

Figure 6. Fossolo network and PAT location.

After the initial design, the TN network was obtained, varying its size from 10% to 100% of the pipes of the network. This procedure was adopted to identify the saturation in resilience and the energy recovery improvement, according to the increase of the TN size. Ten scenarios were explored, as shown in Figure 7, where the red lines represent the TN and blue squares the water source.

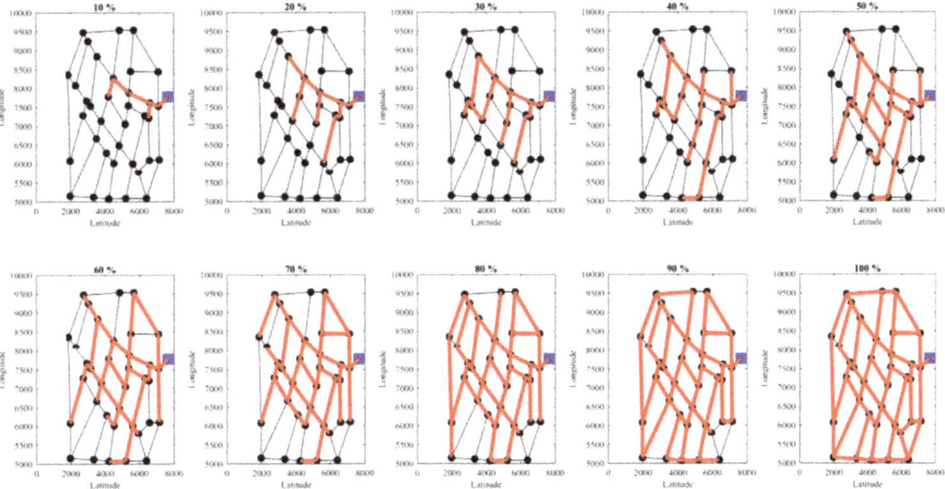

Figure 7. Scenarios for the TN size; red lines represent the TN and blue squares the water source.

For each scenario, the TN is designed again jointly with the PAT selection. The obtained results are shown in Figure 8. It is noticeable that for a TN with 20% of the pipes, both energy production and resilience significantly increase when compared to the initial scenario. These parameters remain approximately constant up to a TN size of 60%, when benefits that were obtained from energy production rise again. This occurs due to the presence of highly elevated nodes, which are located far from the reservoir, and with significant demands. The TN has to be overlong to reach these points. As an alternative, isolated pipes could be enforced separately from a smaller TN. Figure 8d shows the importance of the TN. With small sizes, the increment in diameter is higher, since only important pipes are considered in the design. When the size of the TN increases, less relevant pipes are considered in the design process, and a change in their diameter is irrelevant for the pressure of the network. The results show that approximately three levels of benefit can be distinguished, where additional increments are not important: low (TN with size below 10%), medium (TN with size from 20% to 50%), and high (TN with size above 50%) levels.

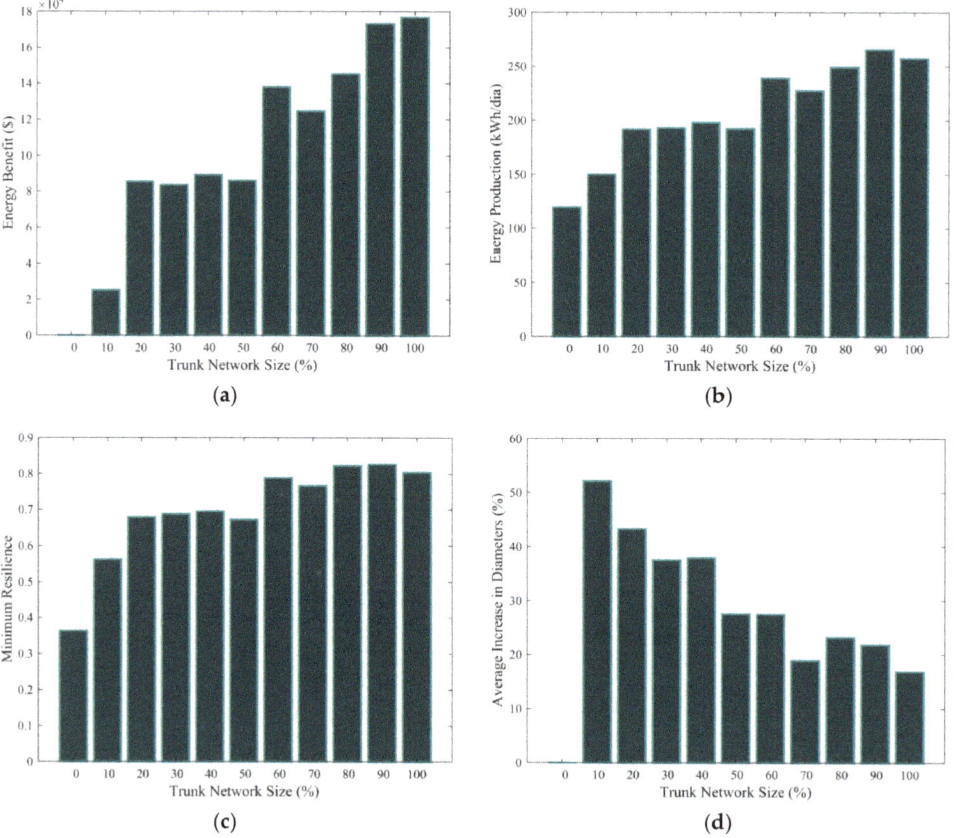

Figure 8. Results of TN rehabilitation for the Fossolo network: (**a**) Energy benefit; (**b**) Energy produced; (**c**) Minimum resilience; and, (**d**) Average increase in pipe diameters.

Another interesting feature to be observed is the behavior of resilience during a 24 h period, as shown in Figure 9. It is observed that even with demand variations, the amplitude of resilience is significantly reduced for TNs with sizes over 20%. Therefore, the reliability of the network is greatly improved, and, even in emergency situations, water supply is not harmed.

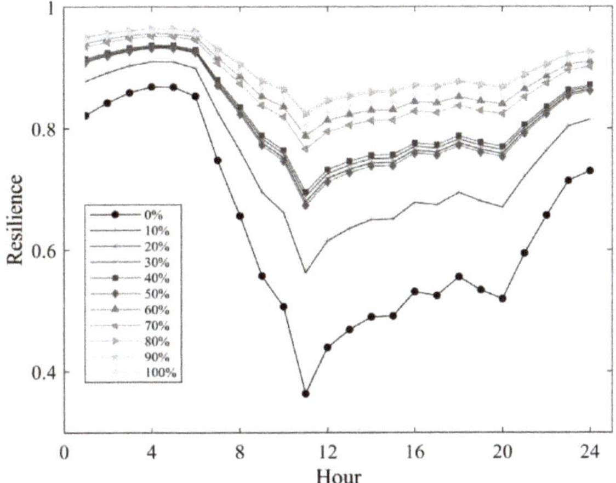

Figure 9. Resilience for a 24 h period in the Fossolo network.

As resilience shows the capacity to meet demands during emergency situations, pipe bursts were simulated. To this end, an emitter coefficient was added to each node to simulate a leak flow that is caused by a pipe burst of 10% of the total inflow. The case of two ruptures occurring at the same time was not simulated. Figure 10 shows the behavior of resilience, according to the size of the TN and the node where the pipe burst occurred. It can be seen that bursts occurring in nodes 20 to 25 are more relevant for system resilience. These nodes are located in a central area of the network, where the TN diameters are gradually reduced, since the flow is also reduced. Thus, an additional flow in this area will significantly increase the headloss of the system. Although pipes that are far from the reservoir have smaller diameters, their flows are very low. So, a demand increase does not increase their headloss too much. On the other hand, pipes close to the reservoir are well sized, and have enough capacity to support an additional demand. Finally, it is noticeable that for TNs above 20% of the pipes, resilience is less affected by a pipe burst. This can be confirmed when comparing Figure 11a, where the pressure zones for the highest consumption period are shown for the initial scenario, and with TNs of 20% and 60% of the pipes with no leakage; and, Figure 11b, where a pipe burst on node 22 occurred. The rehabilitation of the TN significantly increased the pressure in the entire network. In addition, when the pipe burst is simulated, the pressure drop is reduced, and the minimum value of 10 m is maintained, which is not achieved in the standard scenario.

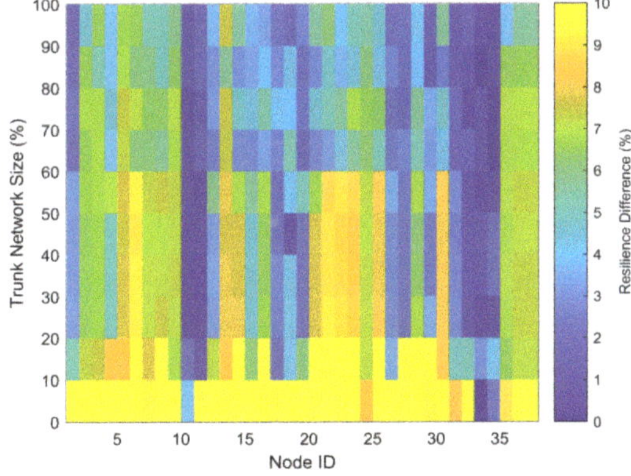

Figure 10. Resilience reduction according to TN size and the location of a pipe burst.

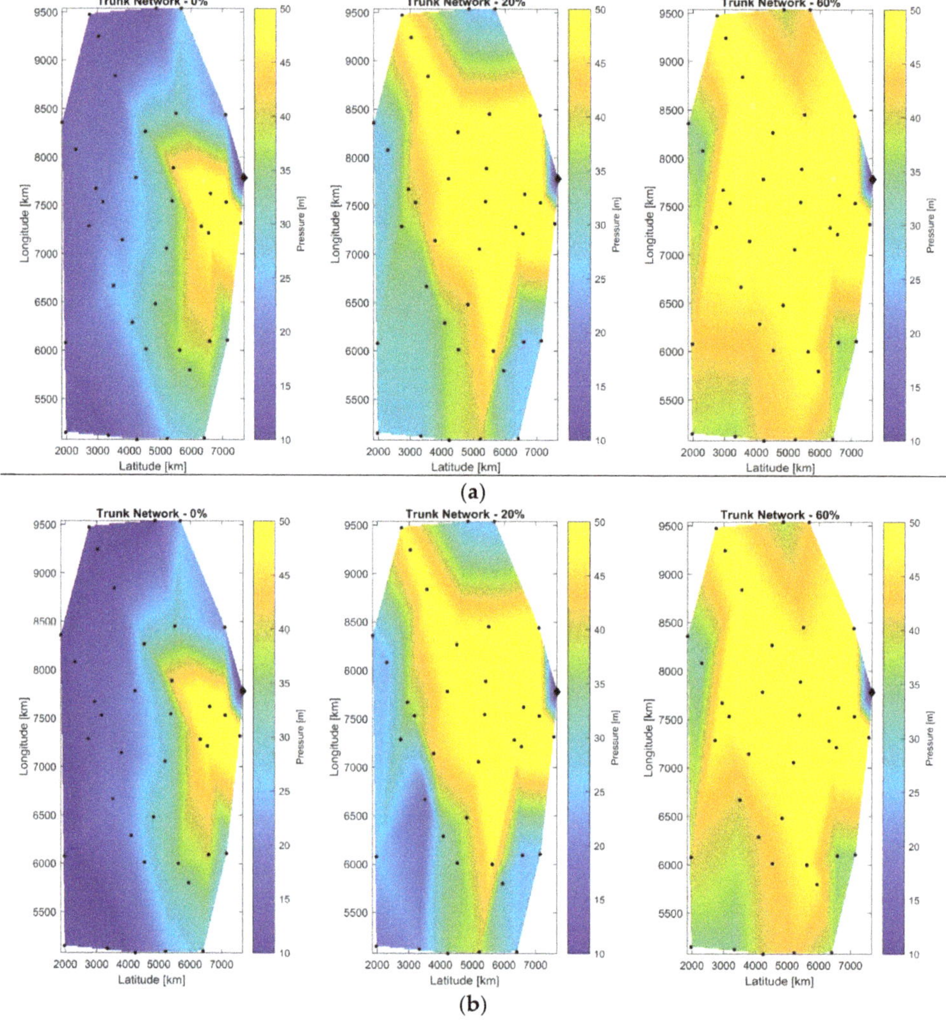

Figure 11. Pressure zones for the initial scenario and for TNs with 20% and 60% of the pipes: (**a**) Without leakage; and, (**b**) Pipe burst on node 22.

3.2. Balerma Irrigation Network

The second case study considers a larger system, the Balerma Irrigation Network [26]. It has 454 pipes and 443 consumption nodes. Four reservoirs supply the system. Therefore, in each of their outlet pipes, a PAT was installed for pressure control and energy recovery, with a total of six machines, as shown in Figure 12. The minimum pressure is also set to 10 m. Based on the results that were obtained in the Fossolo network, three different sizes for the TN were studied: 20%, 60%, and 90% of the total number of pipes, since these were the sizes that exhibited significant parameter improvement. These scenarios are shown in Figure 13, where the red lines represent the TN and the blue squares the water sources.

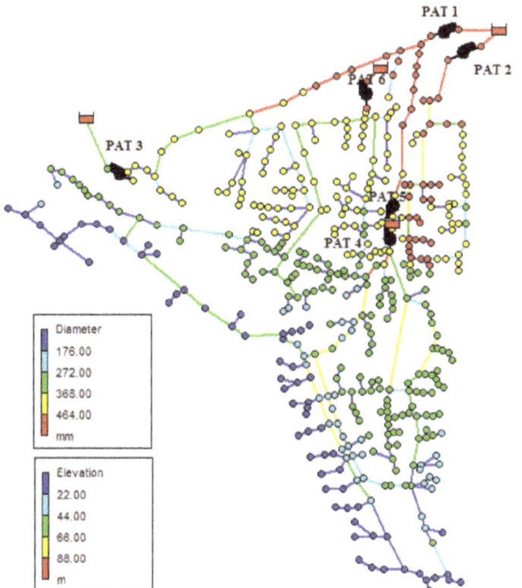

Figure 12. Balerma Irrigation network and PATs location.

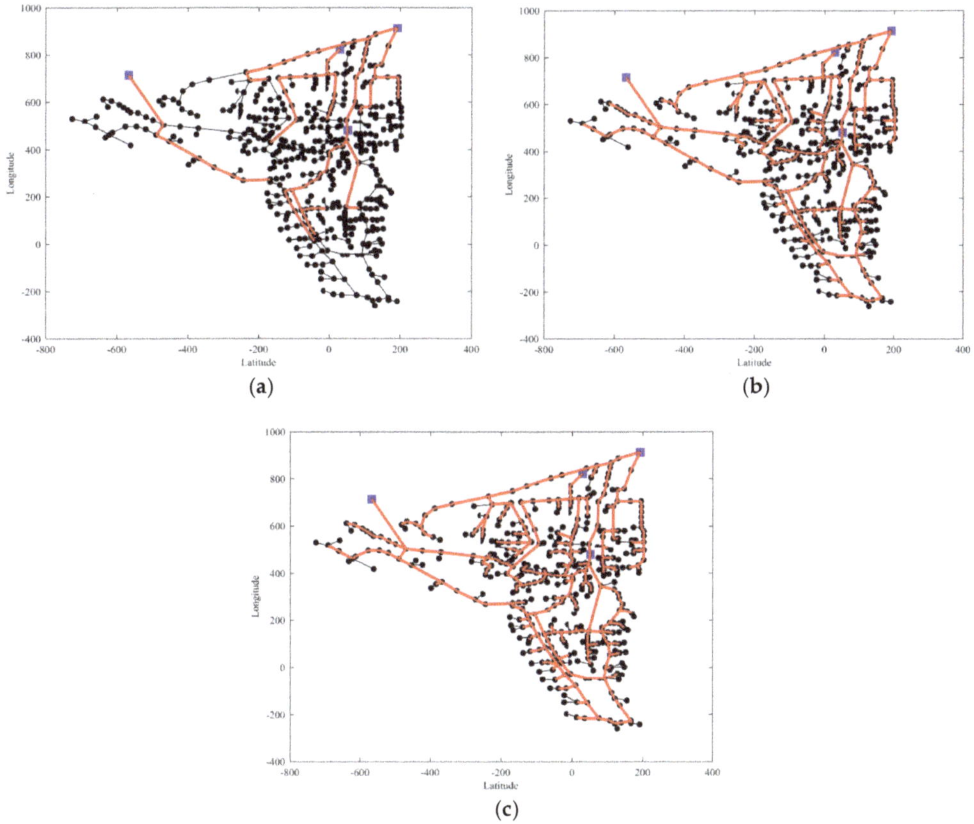

(a)

(b)

(c)

Figure 13. Scenarios for the TN size: (**a**) 20%; (**b**) 60%; and, (**c**) 90%; the red lines represent the TN and the blue squares the water sources.

The results that are presented in Figure 14 show that, for the Balerma case, a TN size increase does not significantly improve energy production and resilience. This occurs due to five consumption nodes that are located near PAT #1. Their high elevations limit the head of PATs #1 and #2, which are responsible for the majority of the flow that is supplied. Thus, the energy production is highly affected by this behavior. However, using a TN with a size of 20% of the pipes is economically feasible, and produce a slightly improvement on network resilience (Figure 14c).

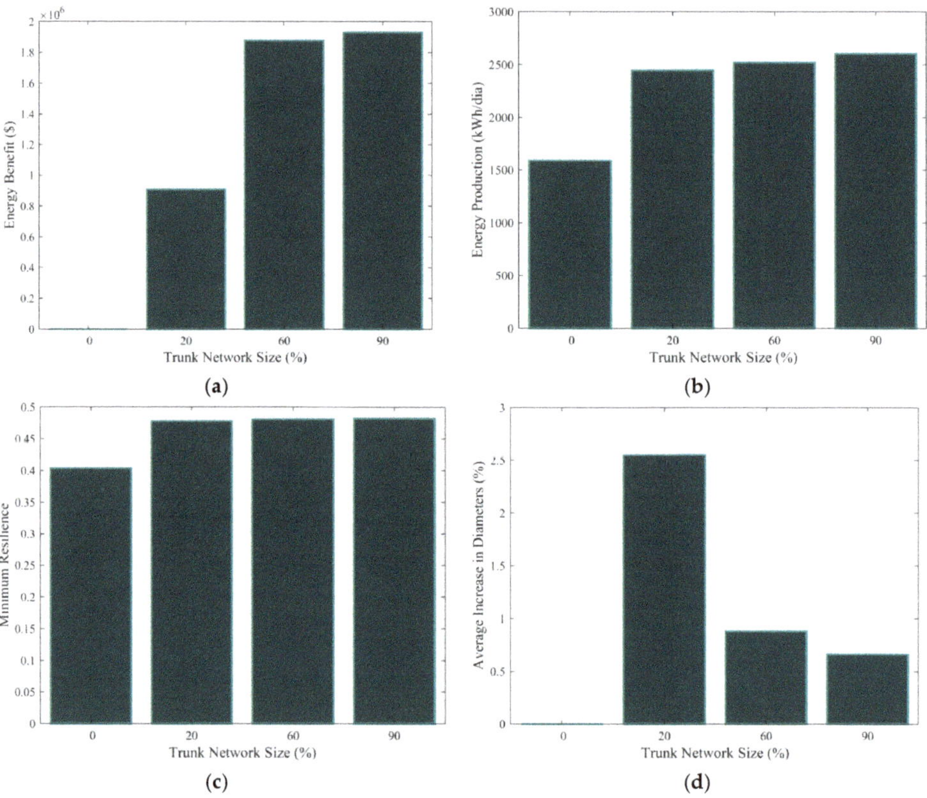

Figure 14. Results of TN rehabilitation for the Balerma irrigation network: (**a**) Energy benefit; (**b**) Energy produced; (**c**) Minimum resilience; and, (**d**) Average increase in pipes diameters.

4. Conclusions

Pipe cost represents the major component in the design of WDNs. However, for long-term operation, minimum investment in pipes can be economically ineffective, since reinforcements can be necessary to meet growing demands. In addition, a system that is capable of operating only with minimum pressure values is not sufficiently robust to surpass emergency situations, such as pipe bursts or firefighting, causing water shortage in some areas. To avoid this problem, WDN resilience is an important parameter to be considered, since it reflects the network capacity to cope with a variety of events. However, it is hard to quantify its importance, since these emergency situations occur occasionally, and it is up to the managers to choose an adequate solution. The methodology that is proposed in this paper explicitly quantifies the benefit of resilience improvement through an energy recovery possibility, stating its importance as a technical and economical parameter. During normal operation, PATs operate to perform pressure control, while, in emergency scenarios, the flow can be driven through a by-pass to regulate pressure at critical points. For existing networks, rehabilitation only of the TN is an alternative to improve resilience, since TN pipes are the most relevant for the system. The case studies presented show the importance of TNs for resilience. While considering a TN of only 20% of the pipes significantly improved the performance of the system. In addition, the energy recovery produced benefits that surpass the additional investment necessary, thus being an attractive alternative for WDN design. Besides, this procedure can be applied to any WDN operating by gravity, and the benefits obtained will depend on its topography and demand, i.e., the available power for energy recovery. In some cases, this approach can lead to unfeasible conditions due to low power availability, and subjective parameters, such as the supply security, have to be considered to value the resilience.

Author Contributions: Conceptualization, G.M., B.B. and E.L.Jr.; Software, G.M.; Investigation, G.M., B.B., E.L.Jr., H.R., J.I.; Writing-Original Draft Preparation, G.M., B.B., E.L.Jr., H.R., J.I.

Acknowledgments: The authors wish to thank the project REDAWN (Reducing Energy Dependency in Atlantic Area Water Networks) EAPA_198/2016 from INTERREG ATLANTIC AREA PROGRAMME 2014–2020.

Conflicts of Interest: The authors declare no conflict of interest.

References

1. Geem, Z.W.; Kim, J.H.; Loganathan, G.V. A New Heuristic Optimization Algorithm: Harmony Search. *Simulation* **2001**, *76*, 60–68. [CrossRef]
2. Maier, H.R.; Simpson, A.R.; Zecchin, A.C.; Foong, W.K.; Phang, K.Y.; Seah, H.Y.; Tan, C.L. Ant colony optimization for design of water distribution systems. *J. Water Res. Plan. Manag.* **2003**, *139*, 200–209. [CrossRef]
3. Suribabu, C.R.; Neelakantan, T.R. Design of water distribution networks using particle swarm optimization. *Urban Water J.* **2006**, *3*, 111–120. [CrossRef]
4. Baños, R.; Reca, J.; Martínez, J.; Gil, C.; Márquez, A.L. Resilience indexes for water distribution network design: A performance analysis under demand uncertainty. *Water Res. Manag.* **2011**, *25*, 2351–2366. [CrossRef]
5. Shokoohi, M.; Tabesh, M.; Nazif, S.; Dini, M. Water quality based multi-objective optimal design of water distribution systems. *Water Res. Manag.* **2017**, *31*, 93–108. [CrossRef]
6. Marques, J.; Cunha, M.; Savić, D. Using real options in the optimal design of water distribution networks. *J. Water Res. Plan. Manag.* **2014**, *141*, 1–10. [CrossRef]
7. Schwartz, R.; Housh, M.; Ostfeld, A. Least-Cost Robust Design Optimization of Water Distribution Systems under Multiple Loading. *J. Water Res. Plan. Manag.* **2016**, *142*. [CrossRef]
8. Giustolisi, O.; Laucelli, D.; Colombo, A.F. Deterministic versus stochastic design of water distribution networks. *J. Water Res. Plan. Manag.* **2009**, *135*, 117–127. [CrossRef]
9. Lansey, K.E.; Duan, N.; Mays, L.W.; Tung, Y.K. Water distribution system design under uncertainties. *J. Water Res. Plan. Manag.* **1989**, *115*, 630–645. [CrossRef]
10. Zheng, F.; Simpson, A.; Zecchin, A. Improving the efficiency of multi-objective evolutionary algorithms through decomposition: An application to water distribution network design. *Environ. Model. Softw.* **2015**, *69*, 240–252. [CrossRef]
11. Geem, Z.W. Multiobjective optimization of water distribution networks using fuzzy theory and harmony search. *Water* **2015**, *7*, 3613–3625. [CrossRef]
12. Prasad, T.D.; Park, N.S. Multiobjective genetic algorithms for design of water distribution networks. *J. Water Res. Plan. Manag.* **2004**, *130*, 73–82. [CrossRef]
13. Ramos, H.M.; Borga, A.; Simão, M. New design solutions for low-power energy production in water pipe systems. *Water Sci. Eng.* **2009**, *2*, 69–84.
14. Carravetta, A.; Houreh, S.D.; Ramos, H.M. *Pumps as Turbines: Fundamentals and Applications*, 1st ed.; Springer: Cham, Switzerland, 2017.
15. Pérez-Sánchez, M.; Sánchez-Romero, F.J.; Ramos, H.M.; López-Jiménez, P.A. Energy recovery in existing water networks: Towards greater sustainability. *Water* **2017**, *9*, 97. [CrossRef]
16. De Marchis, M.; Freni, G. Pump as turbine implementation in a dynamic numerical model: cost analysis for energy recovery in water distribution network. *J. Hydroinform.* **2015**, *17*, 347–360. [CrossRef]

17. Carravetta, A.; del Giudice, G.; Fecarotta, O.; Ramos, H.M. PAT design strategy for energy recovery in water distribution networks by electrical regulation. *Energies* **2013**, *6*, 411–424. [CrossRef]

18. Meirelles, G., Jr.; Luvizotto, E.; Brentan, B.M.; Ramos, H.M. Leakage Control and Energy Recovery Using Variable Speed Pumps as Turbines. *J. Water Res. Plan. Manag.* **2017**, *144*. [CrossRef]

19. Carravetta, A.; Del Giudice, G.; Fecarotta, O.; Ramos, H.M. Energy production in water distribution networks: A PAT design strategy. *Water Res. Manag.* **2012**, *26*, 3947–3959. [CrossRef]

20. Lydon, T.; Coughlan, P.; McNabola, A. Pump-as-turbine: Characterization as an energy recovery device for the water distribution network. *J. Hydraul. Eng.* **2017**, *143*. [CrossRef]

21. Afshar, A.; Jemaa, F.B.; Marino, M.A. Optimization of hydropower plant integration in water supply system. *J. Water Res. Plan. Manag.* **1990**, *116*, 665–675. [CrossRef]

22. Meirelles, G.; Brentan, B.M., Jr.; Luvizotto, E. Optimal design of water supply networks using an energy recovery approach. *Renew. Energy* **2018**, *117*, 404–413. [CrossRef]

23. Campbell, E.; Izquierdo, J.; Montalvo, I.; Ilaya-Ayza, A.; Pérez-García, R.; Tavera, M. A flexible methodology to sectorize water supply networks based on social network theory concepts and multi-objective optimization. *J. Hydroinform.* **2016**, *18*, 62–76. [CrossRef]

24. Di Nardo, A.; Di Natale, M.; Giudicianni, C.; Greco, R.; Santonastaso, G.F. Complex network and fractal theory for the assessment of water distribution network resilience to pipe failures. *Water Sci. Technol. Water Supply* **2018**, *18*, 767–777. [CrossRef]

25. Bragalli, C.; D'Ambrosio, C.; Lee, J.; Lodi, A.; Toth, P. On the optimal design of water distribution networks: A practical MINLP approach. *Optim. Eng.* **2012**, *13*, 219–246. [CrossRef]

26. Reca, J.; Martínez, J. Genetic algorithms for the design of looped irrigation water distribution networks. *Water Resour. Res.* **2006**, *42*, 1–9. [CrossRef]

27. Di Nardo, A.; Di Natale, M.; Santonastaso, G.F.; Tzatchkov, V.G.; Alcocer-Yamanaka, V.H. Water network sectorization based on graph theory and energy performance indices. *J. Water Resour. Plan. Manag.* **2013**, *140*, 620–629. [CrossRef]

28. Hajebi, S.; Temate, S.; Barrett, S.; Clarke, A.; Clarke, S. Water distribution network sectorisation using structural graph partitioning and multi-objective optimization. *Proc. Eng.* **2014**, *89*, 1144–1151. [CrossRef]

29. Moore, E.F. *The Shortest Path through a Maze*; Part II: The Annals of the Computation Laboratory of Harvard University Volume XXX; Harvard University Press: Cambridge, MA, USA, 1959; pp. 285–292.

30. Todini, E. Looped water distribution networks design using a resilience index based heuristic approach. *Urban Water* **2000**, *2*, 115–122. [CrossRef]

31. Eberhart, R.; Kennedy, J. A New Optimizer Using Particle Swarm Theory. In Proceedings of the Sixth International Symposium on Micro Machine and Human Science, 1995 (MHS '95), Nagoya, Japan, 4–6 October 1995; pp. 39–43.

32. Brentan, B.M.; Campbell, E.; Meirelles, G.L.; Luvizotto, E.; Izquierdo, J. Social Network Community Detection for DMA Creation: Criteria Analysis through Multilevel Optimization. *Math. Probl. Eng.* **2017**, *2017*. [CrossRef]

33. Meirelles, G., Jr.; Luvizotto, E.; Brentan, B.M. Selection and location of Pumps as Turbines substituting pressure reducing valves. *Renew. Energy* **2017**, *109*, 392–405.

34. Letting, L.K.; Hamam, Y.; Abu-Mahfouz, A.M. Estimation of Water Demand in Water Distribution Systems Using Particle Swarm Optimization. *Water* **2017**, *9*, 593. [CrossRef]

35. Rossman, A.L. *EPANET 2.0 User's Manual*; EPA/600/R-00/057; U.S. Environmental Protection Agency: Cincinnati, OH, USA, 2000.

36. ABNT–Brazilian Association of Technical Standards. *Project of Water Distribution Network for Public Supply*; NBR 12218; ABNT–Brazilian Association of Technical Standards: Rio de Janeiro, Brazil, 1994. (In Portuguese)

© 2018 by the authors. Licensee MDPI, Basel, Switzerland. This article is an open access article distributed under the terms and conditions of the Creative Commons Attribution (CC BY) license (http://creativecommons.org/licenses/by/4.0/).

Article

Validation of a Computational Fluid Dynamics Model for a Novel Residence Time Distribution Analysis in Mixing at Cross-Junctions

Daniel Hernández-Cervantes [1], Xitlali Delgado-Galván [2], José L. Nava [2], P. Amparo López-Jiménez [3], Mario Rosales [1] and Jesús Mora Rodríguez [2,*]

[1]   Ph.D. Student of Doctoral program on Water Sciences and Technology, Engineering Division, Universidad de Guanajuato, Av. Juárez No. 77, Centro, Guanajuato 36000, Mexico; d.hernandezcervantes@ugto.mx (D.H.-C.); m.rosalesretana@ugto.mx (M.R.)
[2]   Geomatics and Hydraulics Engineering Department, Universidad de Guanajuato, Av. Juárez No. 77, Centro, Guanajuato 36000, Mexico; xdelgado@ugto.mx (X.D.-G.); jlnm@ugto.mx (J.L.N.)
[3]   Hydraulic Engineering and Environment Department, Universitat Politècnica de València, Camino de Vera, s/n, 46022 Valencia, Spain; palopez@upv.es
*   Correspondence: jesusmora@ugto.mx; Tel.: +52-473-102-0100

Received: 30 April 2018; Accepted: 28 May 2018; Published: 5 June 2018

Abstract: In Water Distribution Networks, the chlorine control is feasible with the use of water quality simulation codes. EPANET is a broad domain software and several commercial computer software packages base their models on its methodology. However, EPANET assumes that the solute mixing at cross-junctions is "complete and instantaneous". Several authors have questioned this model. In this paper, experimental tests are developed while using Copper Sulphate as tracer at different operating conditions, like those of real water distribution networks, in order to obtain the Residence Time Distribution and its behavior in the mixing as a novel analysis for the cross-junctions. Validation tests are developed in Computational Fluid Dynamics, following the k-ε turbulence model. It is verified that the mixing phenomenon is dominated by convection, analyzing variation of Turbulent Schmidt Number vs. experimental tests. Having more accurate mixing models will improve the water quality simulations to have an appropriate control for chlorine and possible contaminants in water distribution networks.

Keywords: water distribution networks; EPANET; safe water

---

1. Introduction

In recent years, useful life of a several number of Water Distribution Networks (WDN) has been exceeded, so that there are certain points or sectors that do not reach enough disinfectant. This problem could produce the formation of biofilm on pipe walls [1,2], promoting bacterial growth and its transport through water. On the other hand, the system's own useful life, the intermittency in the supply, the leaks level, and the demands behaviour cause a great variety of residence times of the drinking water and variations of reaction of the disinfectant in the network [3,4].

Chlorine disinfection is one of the principal factors in drinking water treatment processes. This chemical agent is used mainly to ensure the destruction of pathogen organisms that could be present in water and that are coming from the sources of supply. The disinfection prevents gastrointestinal diseases due to the consumption of contaminated water. However, it is also true that chemical disinfection has caused unwanted risks due to by-products that are generated in the reactions with contaminants present in water, such as trihalomethanes (THM), among others [1,2,5–9].

Finding solutions to the problem of inadequate use of chlorine is an issue that involves various aspects. When WDN presents a high degree of reaction due to pipe conditions and residence times, concentrations in some areas decrease rapidly [2,3,7]. Due to this, the absence of the necessary amount of disinfectant induces the possible consumption of water that is contaminated by users. A commonly used solution is the increased chlorine concentration from the sources. However, demand nodes near the sources distribute water with the high chlorine contents and due to factors, such as residence times, the formation of harmful by-products increases, which in the same way affects the consumer population [9,10].

Location of strategic points to control the problem of some solute distribution can be reached through the applications, in which it is necessary to improve the incomplete mixing model. Diverse authors [11–14] have developed models based on heuristic technics that use the results of water quality analysis to determine the best location of booster chlorination stations to optimize their usage. Other authors [15–20] developed researches that determine the strategic zones for the sensors location to detect pollutants, with the aim of finding accidental or intentional action of harmful agents and with the idea that the consumer has the risk of consuming contaminated water. Both kinds of applications predict the best zones for the water quality control devices; however, they are dependent on the way, in which EPANET (is a computer program that performs extended period simulation of hydraulic and water quality behavior within pressurized pipe networks [21]. The name comes from the U.S. Environmental Protection Agency and Networks) simulates the distribution of a solution through the network; the predictions are not accurate as many cross-junctions are presented in the networks that are studied.

An incomplete mixing analysis is presented in experimental tests that use injection pulses of copper sulphate ($CuSO_4$) as a tracer. Residence Time Distribution (RTD) curves are good features and a novel proposal on mixing cross-junctions, in order to calibrate simulation processes, with the help of stimulus-response techniques [22–24]. There were generated turbulent regime scenarios in Reynolds numbers from 43,343 to 155,793. Therefore, different operating conditions that are close to real WDN were performed. To analyse the tracer distribution in the outlets of the cross-junction, the curves of the pulses are registered, visualizing by RTD curves the approximation of the Computational Fluid Dynamics (CFD) simulations. According to the cross-junction mixing models that are reviewed in the literature, this is the first time that the hydraulic pipe flow has been simulated with success using RTD analysis, which was addressed by solving the RANS equations with the k-epsilon turbulence model coupled to the diffusion-convection equation.

Complete Mixing Model

In EPANET and in most hydraulic simulation programs, it is assumed that the solute mixing is complete and instantaneous [21]. Therefore, the concentration of the substance (for example: residual chlorine) that leaves the cross-junction, will be the weighted average of the solute concentrations of the arriving pipes. For a given node k, the concentration leaving the node is obtained by Equation (1).

$$C_i = \frac{\sum_{j\in I_k} Q_j C_j + Q_{k,ext} C_{k,ext}}{\sum_{j\in I_k} Q_j + Q_{k,ext}} \tag{1}$$

where: i, link with flow leaving node k; I_k, set of links with flow into k; L_j, length of link; Q_j, flow (volume/time) in link j; $Q_{k,ext}$, external source flow entering the network at node k; and, $C_{k,\,ext}$, concentration of the external source flow entering at node k.

Several authors [25–36] have numerically and experimentally demonstrated that the mixing in these unions is far from being "complete and instantaneous". Most of these researches were based on physical experiments and simulations through Computational Fluid Dynamics (CFD) using tracers for the concentration distribution analysis. Some researchers [26–28,31–36] implement experimental scale models of a cross-junction and tee junctions with different pipe diameters, which are driven by elevated tanks where tracer is injected into one of them. Once the mix of the flows was made, mixing coefficients were formulated based on their experimental measurements. Some others based their approaches on CFD simulations directly, even for small geometric configurations [25,29,30]. The most commonly used tracers were salts, such as NaCl. Other authors considered colored dyes and high speed photography too [34,35]. Conducting a study directly with chlorine as a solute turns out to be somewhat complicated, although Mompremier [33] used in-line chlorine loggers/dosers to analyze the mixing at different conditions.

The developed studies carried out to date allow for the derivation of mathematical functions for the computation of the incomplete mixing, even two codes have been formulated to be implemented as toolkits for EPANET (one of them, called AZRED II and developed by [36]). However, these applications have not been disseminated globally yet, because most of them were performed by scaled experimental devices reaching low velocities and turbulent flows that were close to 4000. In some cases, they are still subject to specific calibrations of each network and there is a lack of experimental sufficiency in real networks of high interconnection.

The lack of precision in water quality simulations can lead to erroneous projects of monitoring systems of the location of re-chlorination systems necessary for the control of disinfection [28]. The accuracy in these models is also necessary to simulate the spatial-temporal dispersion of chemical and microbial agents during accidental or intentional contamination events.

This paper presents a validation of the phenomenon of mixing in cross-junctions by means of CFD, in which the distribution of the disinfectant is not considered as homogenous mixing. The presented scenarios were performed at velocity and pressure conditions that were similar to those real WDN thanks to the impulsion and conduction equipment used. This avoids making measurement adjusts or extrapolations, that makes it different from many other jobs working at lower turbulence flows. The model is validated by a novel analysis of RTD curves for mixing cross-junction, resulting from the execution of tests of different velocity conditions and its corresponding measurement of copper sulfate injection response. The correct model meshing degree is verified for the study of the phenomenon in the CFD model.

2. Materials and Methods

2.1. Experimental Model

The cross-junction used for the mixing analysis is situated on an experimental network of the "Universidad de Guanajuato"; the pipe material is galvanized iron of four inches in diameter (Figure 1). The flows are conducted by a 15HP hydraulic pump, which reaches 30 L per second at pressures of 1.78 bar for the pipes on the cross-junction. Open valves are available at each border for flow velocity control for inlets and outlets. They are located at 2 and 7.5 m upstream water from the cross-junction. It is done to reduce turbulence generated by the obstruction of the valve, and so that the flow could be uniformized along the pipeline.

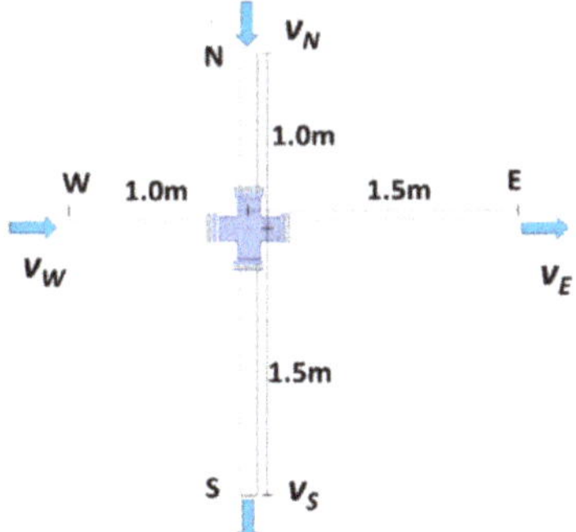

Figure 1. Cross-junction boundaries. The distances correspond to the pressure registration points.

Operating conditions that are close to WDN can be approximated by the physical characteristics of the experimental network. All of the mixing tests were developed contemplating two inlets and two outlets flow. The boundary distances (shown in Figure 1), that were established for the numerical model, are the location points of the pressure measuring devices. The north inlet (N) corresponds to the upper boundary. The West inlet (W) is the left boundary, and in the same way for the outlets boundaries on the East (E) and South (S) (Figure 1). In addition, the numerical model is performed with the objective of maintaining the flow direction with a sense from left to right and from top to bottom.

North and West inlets flows are measured with the support of two flow meters, of them being of electromagnetic type, DOROT brand, model MC608-A (Figure 2a). The second one is a propeller meter, of the company Hidrónica®, model MP-0400 (Figure 2b), both with datalogger capabilities. For measures at the outlets, the outflowing water is conducted up to four flowmeters, two for each outlet. For the East outlet, the flowmeters are of electromagnetic type, of Badger Meter Inc. brand, model M2000 (Figure 2c). For the South outlet, the water is conducted to two propeller volumetric flow meters (Figure 2d).

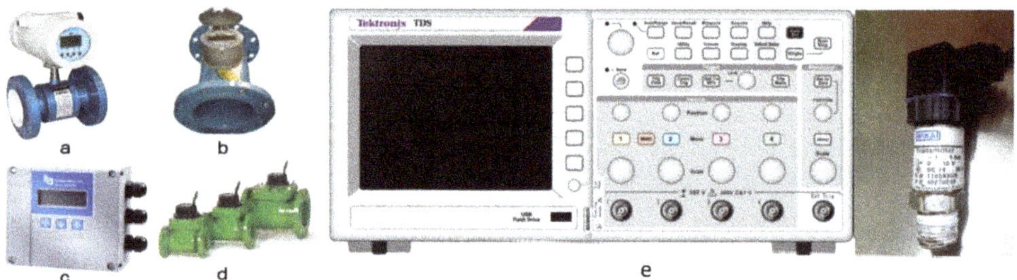

a
b
c
d
e

Figure 2. Equipment to measure flow (electromagnetic: (**a**,**c**), propeller volumetric: (**b**,**d**)) and pressure around the cross-junction (oscilloscope and pressure transducer: (**e**)).

A digital storage oscilloscope with four measuring channels, a Tektronix® brand model TDS-2004C (Figure 2e) was available to the pressure registration on the four boundaries of the cross-junction. In this case, the oscilloscope visually represents the electrical signals that are captured by pressure transducers. These transducers are of WIKA® brand model S-10, comprise a measuring range of 0–10 V of electrical signal intensity corresponding to the range -1 to 9 bar of pressure.

The tracer that was used for the tests is copper sulphate pentahydrate ($CuSO_4 \cdot 5H_2O$) of fine grade, Fermont® brand. The 0.25 mol/L solutions are prepared, which return an effective response to the concentration measurement. The tracer injection is done by pumping, with a monophasic equipment, Siemens brand, 1HP power and rotor speed of 3515–3535 rpm. The tracer is stored in a volumetric test tube of four liters size, and then, in each test, the tracer is pumped 150 mL approximately. The injection point is located at 3.50 m before the cross-junction (Figure 3). The pump equipment allows for overcoming the pressures of water flow in the network and the controlled intrusion of tracer that is injected through the opening and closing ball valve of a 0.5 inch.

A potentiostat-galvanostat, brand BIOLOGIC, model SP-150 was used to apply an electric potential, with which copper adhesion will occur on the electrodes (Figure 4). This equipment has software that allows for current intensity registration at fixed time intervals. The electrodes were placed at the outlet boundaries of the pipes, close to the pressure transducers. The electrodes must be made of a metal with suitable properties for electrical conduction. They were developed with the characteristics that are shown in the Table 1.

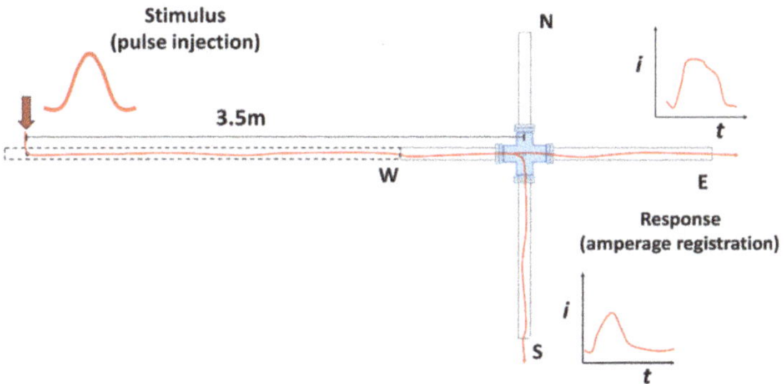

Figure 3. Tracer injection and registration points.

Figure 4. Equipment for stimulus-response technique.

Table 1. Characteristics of the electrodes.

Material	Solid Copper
Electrode diameter	4.4 mm
Length of electrodes	9.5 cm (covers more than 90% of the pipe diameter)
Electrodes separation	4.6 mm
Insulation material	resin

The velocity alternatives for the case of two inlets and two outlets are described in Figure 5. To simulate equal velocities is difficult because of working with high magnitudes of velocity and pressure. Therefore, the scenarios were approximated as much as possible to cover the four-velocity combinations in Figure 5. It has been determined that the flow velocities are closer when the variation between both of the inlets or outlets boundaries are by a percentage with a relative difference is less than 25%, otherwise, for the variations that are presented with a relative difference of 25% or greater, it is considered that the velocities are different. The objective is to cover trials that involve the different mixing conditions, likely in a cross joint. The alternatives are classified in these four groups, according to their velocities. Table 2 contains contents values of the four scenarios V1 to V4 proposed.

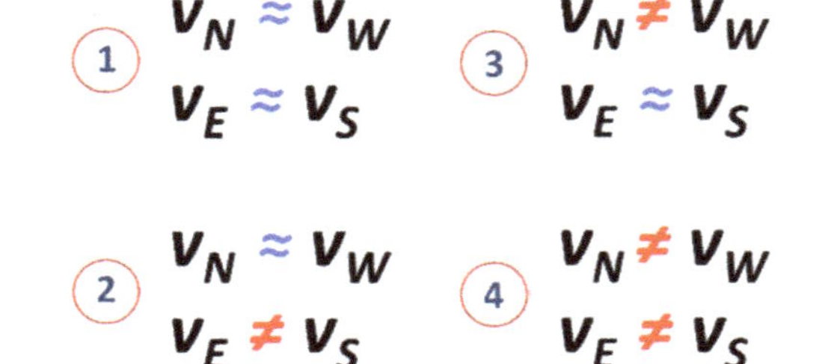

Figure 5. Boundary velocities for experiments.

Table 2. Contents values of the four scenarios V1 to V4 proposed.

Front		V1				V2				V3				V4			
		V (m/s)	P (bar)	Re	Q (l/s)	V (m/s)	P (bar)	Re	Q (l/s)	V (m/s)	P (bar)	Re	Q (l/s)	V (m/s)	P (bar)	Re	Q (l/s)
Inlets	N	1.112	1.88	112,928	9.01	1.116	1.60	113,396	9.05	0.808	1.57	82,123	6.55	0.672	1.96	68,295	5.45
	W	1.059	1.88	107,625	8.59	1.269	1.60	128,971	10.29	1.533	1.56	155,793	12.43	0.989	1.96	100,503	8.02
Outlets	E	1.005	1.88	102,067	8.14	1.062	1.56	107,899	8.61	1.039	1.56	105,654	8.43	1.246	1.96	126,624	10.10
	S	1.167	1.88	118,547	9.46	1.324	1.56	134,468	10.73	1.302	1.56	132,273	10.55	0.427	1.96	43,343	3.46

2.2. Formulation of Numerical Simulation

2.2.1. Turbulent Flow

The numerical simulations for the dynamic flow are based on the mass conservation Equation (2) and the Reynolds Averaged Navier-Stokes (RANS). These last equations consider the turbulent flow applying Reynolds decomposition by its average over a small-time increment; the flow quantities are decomposed in a temporal mean and a fluctuating component [37]. The decomposition results in the RANS Equation (3) [38,39].

$$\nabla \cdot (\rho u) = 0 \tag{2}$$

$$\rho(u \cdot \nabla)u = -\nabla P + \nabla \cdot \left((\mu + \mu_T)\left(\nabla \cdot u + (\nabla \cdot u)^T \right) \right) \tag{3}$$

where u is the mean-averaged velocity vector, P is the pressure, μ is the dynamic viscosity, and ρ is the fluid density.

The fluctuating component introduces a closure problem [37], which could be solved by diverse type of models: Eddy Viscosity or First-Order and Second-Moment Closure (SMC). The standard k-ε model is one of SMC that works with high Reynolds number turbulence valid only for fully turbulent free shear flows that cannot be integrated all the way to the wall [40]. Modeling flows close to solid walls requires integration of the two equations over a fine grid to correctly capture the turbulent quantities inside the boundary layer [40,41]. Then, the so-called Reynolds stresses can be stated in terms of a turbulent viscosity μ_T, according to the standard k-ε turbulence model; Equations (4) to (6).

$$u_T = \rho C_\mu \frac{k^2}{\varepsilon} \tag{4}$$

$$\rho(u\nabla)k = \nabla \cdot \left(\left(\mu + \frac{\mu_T}{\sigma_k} \right) \nabla k \right) + P_k - \rho \varepsilon \tag{5}$$

$$\rho u \cdot \nabla \varepsilon = \nabla \cdot \left(\left(\mu + \frac{\mu_T}{\sigma_\varepsilon} \right) \nabla \varepsilon \right) + C_{e1} \frac{\varepsilon}{\kappa} P_k - C_{e2} \rho \frac{\varepsilon^2}{k} \tag{6}$$

where k is the turbulent kinetic energy, ε is the turbulent energy dissipation rate, P_k is the energy production term ($P_k = \mu_T[\nabla u : (\nabla u + (\nabla u)^T)]$), and C_μ (0.09), C_{e1} (1.44), C_{e2} (1.92), σ_k (1), and σ_ε (1.3) are dimensionless constant values that are obtained by data fitting over a wide range of turbulent flows [38]. K. Hanjalik [41] used the k-ε model over other types of turbulence models for simulation in pressure pipes with high Reynolds numbers, with the same empirical constants. Furthermore, RANS k-ε models have proposed good results also in mixing and concentration analysis, when these models have been validated with measurements, as in [42,43].

2.2.2. Boundary Conditions

Due to the high values of the Reynolds number that were reached in the scenarios, velocities in regions near the walls model decrease rapidly. The fluid velocity becomes dependent on these boundaries. The wall functions used are based on a universal distribution of velocities depending on the proximity of the wall, Equation (7).

$$u^+ = \frac{1}{\kappa} \ln\left(E y^+ \right) \tag{7}$$

where u^+ is the normalized velocity component in the logarithmic layer of the wall, κ is the Von Karman constant (0.4187), E is a constant that depends on the roughness of the wall, and y^+ is the dimensionless distance from the wall.

The input parameters that are assigned to the CFD analysis (shown in Table 2) were established them in the following way, at the same distance like in the experimental model:

Inlet velocity at N;

Inlet velocity at W;

Pressure at E outlet; and,

Pressure at S outlet.

The velocity values (Table 2) are multiplied by the unit normal vector n, $u = -vn$, then, the input velocity is a vector of equal magnitude in the surface input boundary. The parameters k_0 and ε_0 must be defined, which are obtained from the Equations (8) and (9).

$$k_0 = \frac{3}{2}(v_0 I_T)^2 \tag{8}$$

$$\varepsilon_0 = \frac{C_\mu^{\frac{3}{4}} k^{\frac{3}{2}}}{L_T} \tag{9}$$

Turbulent intensity I_T (dimensionless) varies between 0.05 and 0.10. For the turbulent scale length, L_T (long unit) can be obtained according to the pipe radius r, obtaining 7% of it. $L_T = 0.07r$ [39]. Therefore, the values were fixed on $I_T = 0.05$ and $L_T = 0.07 \times 0.0508$ m $= 0.003556$ m.

2.2.3. Residence Time Distribution

To sketch RTD curve in a fluid for transport of diluted species in dynamic regime, an averaged convection-diffusion equation for turbulent flow is used, Equation (10).

$$\frac{\partial C}{\partial t} = -\mathbf{u} \cdot \nabla C + \nabla \cdot (D + D_T) \nabla C \tag{10}$$

where u is the velocity vector determined from Equation (2) of the hydrodynamic analysis, C is the tracer concentration, and D is the diffusion coefficient of the tracer. Under turbulent flow conditions, the concentration and the fluctuation parameters can be grouped in a turbulent diffusion term, D_T. This term can be evaluated considering the Schmidt Number (turbulent), Sc_T, for which, the equation proposed by Kays-Crawford [41] was used, Equations (11) and (12).

$$Sc_T = \frac{\mu_T}{(\rho D_T)} \tag{11}$$

$$Sc_T = \frac{1}{\frac{1}{2Sc_{T\infty}} + \frac{0.3\mu_T}{\sqrt{Sc_{T\infty}}\rho D} - \left(0.3\frac{\mu_T}{\rho D}\right)^2 \left(1 - exp\left(-\frac{\rho D}{0.3\mu_T\sqrt{Sc_{T\infty}}}\right)\right)} \tag{12}$$

where $Sc_{T\infty}$ is the value of Sc_T at a distance far from the wall, and it is fixed with a value $Sc_{T\infty} = 0.85$. It is considered to be a perfect mix at the tracer input boundary due to the distance with which it is injected.

The RTD curve is estimated by Equation (13) and it describes the distribution of the tracer through model at certain fixed time periods, from the injection of the tracer to the point of contact with the electrodes at outlets.

$$E(t) = \frac{C(t)}{\int_0^\infty C(t)dt} \tag{13}$$

where $C(t)$ is the concentration that is time-dependent on time. The development of RTD curves was executed taking $C(t)$ from the Equation (10).

Both DTR curves obtained, experimental and simulated will have different dimensions, so it is necessary to make a parameterization. A normalized function $E(\theta)$ can be defined as [23]:

$$E(\theta) = \tau E(t) \tag{14}$$

where the dimensionless time θ is obtained by the spatial residence time $\tau = V/Q$, V is the fluid volume of the model, and Q is the volumetric flowrate in each outlet.

The boundary and the initial conditions for solving the Equation (10) can be established as the tracer concentration, that is zero, $C = 0$, before the tracer injection in the reactor ($t = 0$) is performed. To simulate the tracer pulse injection, a mean form of the RTD is employed RTD at the outlets.

The mixing analysis was implemented in CFD, by the COMSOL Multiphysic software, 4.3b version in a computer of 64-bits operating system, Dual Core Xeon 2.30 GHz processor, and 96 GB of RAM. COMSOL has efficient simulation capabilities and complete graphic settings over other simulation commercial software packages [44]. This allows for better describing of different development patterns by function mathematical assignments in the simulation of tracer injection and the calculation of turbulent flow implemented in this study.

The discretization of the cross-junction volume was done using an unstructured mesh with tetrahedral, pyramid, and prism elements for simulation. According to the COMSOL manual [45], the mesh depends on the fluid flow and on the accuracy required. In many cases, the standard k-epsilon model with wall functions, that in this case, may deliver an accurate enough result at a much lower computational cost.

3. Results

3.1. Mesh Selection and Sensibility Analysis

The COMSOL software already has predefined mesh grades for simulations in turbulent flow, varying only the mesh density degree. A sensitivity analysis (Figure 6) was carried out among nine different fineness degrees to determine which density is adequate for the numeric representation regarding the measurements that were made of pressure and velocity in the real experiment. Also, it is verified that by means of the value of y^+ there is an adequate representation of the velocity profile.

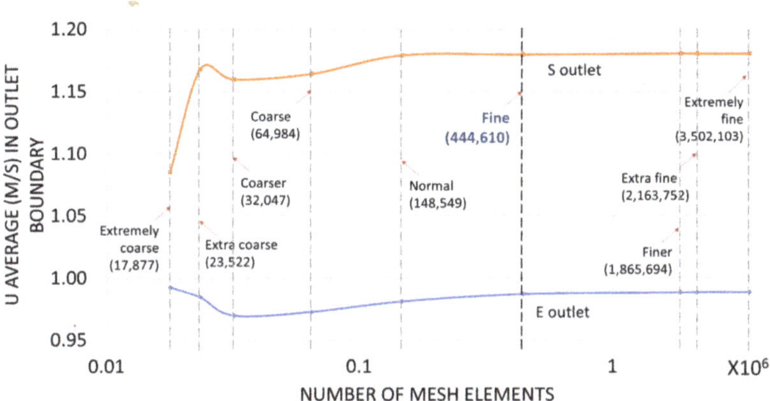

Figure 6. Mesh sensibility analysis.

The sensitivity analysis for scenario V1 is shown in Figure 6. For the degrees of coarse mesh (extremely coarse, extra coarser, coarser, coarse, and normal), the speed is having a variation for both outlets. However, from the FINE fineness level, besides the speeds adjusted and the increase of the degree of fineness, there is no other significant change. Therefore, the appropriate degree of mesh for the simulations is the FINE grade, it is conformed of 493,438 to 444,610 mesh elements, with the minimum and maximum edge lengths of 1.92, 0.362 cm, respectively; the number of elements depends on the drop of pressure that is generated from the experimental valves that control the flow that in the CFD was represented with an obstruction on the flow. Although only one case of sensitivity analysis is shown, it is not necessary to perform one for each scenario, since the geometry of the model varies insignificantly.

According to Veerstek & Malalasekera [38] and Moukalled, Margani & Darwish [46], y^+ values in a range of (11.06 to 12.00) have important effects to turbulence production near the wall. As shown in the graph of Figure 6, the u velocity with respect to the mesh degree (by number of elements) stops varying from the FINE mesh degree. Therefore, this mesh degree was selected for the CFD model, which guarantees that the mesh tetrahedra established in the contours maintains a minimum y^+ value of 11.1 (those were verified plotting values of y^+ in a mesh graphic). The wall roughness was assumed to have a negligible effect. The solver that was employed was iterative, a generalized minimal residual, and a relative tolerance of accuracy of the CFD simulations considered a convergence criterion below 1×10^{-5}. The typical solution around these meshes elements was unchanged.

3.2. CFD Simulation

The tracer injection must be defined in CFD simulations. However, the internal geometry of the sphere valve for the injection is controlled manually, and because of that, it is very irregular and difficult to obtain a chemical input pattern. Therefore, an interpolation was made between the tracer curves obtained at the outputs of the cross-junction. With this, a tracer input average is obtained at the injection point, and with this, the simulation is performed.

In the Table 3, it can be verified that the approximation between the experimental and CFD results has a significant proximity. For velocity values, the average error is 0.76% and the maximum value reached is 1.72%. The errors in terms of pressure are on average 0.80% on the N boundary and 0.92% on the W boundary. In part, pressure differences that are presented will have to do with the decimals with which the readings were recorded. The values in CFD have 16 decimals of floating point, while the oscilloscope used registers of an average value of signal, in a given time, with two decimals.

Table 3. Approximation errors between experimental and simulated tests.

		V1			V2			V3			V4		
		Exp	CFD	% Error	Exp	CFD	% Error	Exp	CFD	% Error	Exp	CFD	% Error
P (bar)	N	1.88	1.893	0.670%	1.60	1.577	1.437%	1.57	1.578	0.502%	1.96	1.971	0.575%
	W	1.88	1.893	0.677%	1.60	1.576	1.478%	1.56	1.575	0.982%	1.96	1.971	0.557%
V (m/s)	E	1.005	0.987	1.720%	1.062	1.060	0.211%	1.040	1.044	0.381%	1.246	1.237	0.744%
	S	1.167	1.180	1.130%	1.324	1.322	0.133%	1.302	1.294	0.593%	0.427	0.422	1.192%

The velocity values that were obtained in CFD correspond to vectors of a velocity field solved in the Navier Stokes Equation (3) for turbulent flow (k-ε model). However, to obtain an average value, COMSOL can estimate averages in a specific area (Figure 7). The selected mesh (FINE grade) approached the area of a contour (the outlet) of the pipe with good approximation (0.641% with respect to the real geometry obtained by a circumference).

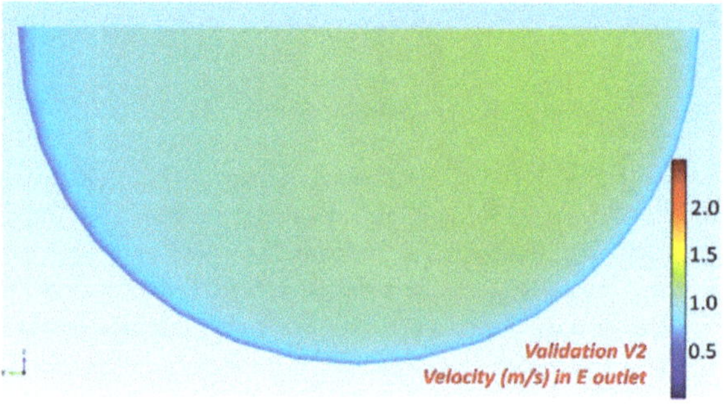

Figure 7. Example of a velocity contour plot.

3.3. Residence Time Distribution Curves

Figures 8–11 show the RTD curves, experimental and numerical, at different inlet velocities (V1 to V4). The current intensity value (*milliAmperes*), recorded in time intervals, varies by the exchange of copper ions in the electrode cells for the experimental RTD. On the other hand, the numerical RTD is the tracer concentration through time. The parameterization was made according to the Equation (14). It is important to emphasize that the CFD curves maintain a constant value most of the time, which is the stabilization of the received signal giving the supplied potential. Once the curves of the experimental and simulated tests are derived, a comparison is made between the curves to analyze the representation of the simulation. The dimensions of the axes of the graphs are: The dimensionless time (θ) on the horizontal axe and the normalized function for the concentration E (θ) on the vertical axe; in all the cases, there are diverse values indicating a not complete mixing. Close agreement between simulation and experimental tests were attained. The shape of the RTD curves evidences that the mixture in the pipe junctions does not obey a perfectly mixed model, owing to the fact that the maximum of the peak is far from θ = 0. The maximum of the curves was found at values of θ, higher than of the unity, what indicates that the fluid elements loss kinetic energy in the cross-junction due to the coalition of the hydrodynamic streams and the bifurcation of the fluid flow.

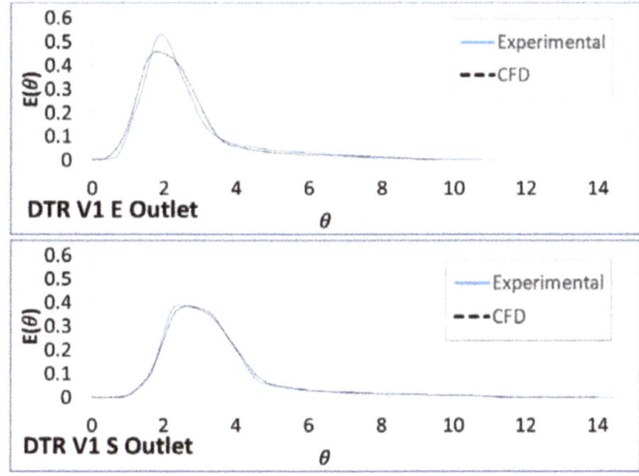

Figure 8. Residence Time Distribution (RTD) curves for V1.

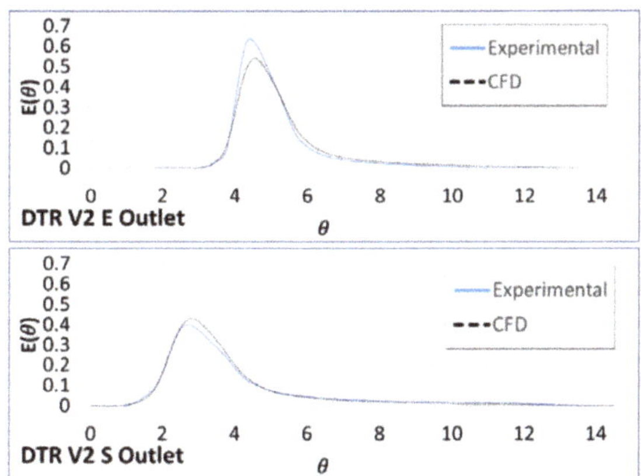

Figure 9. RTD curves for V2.

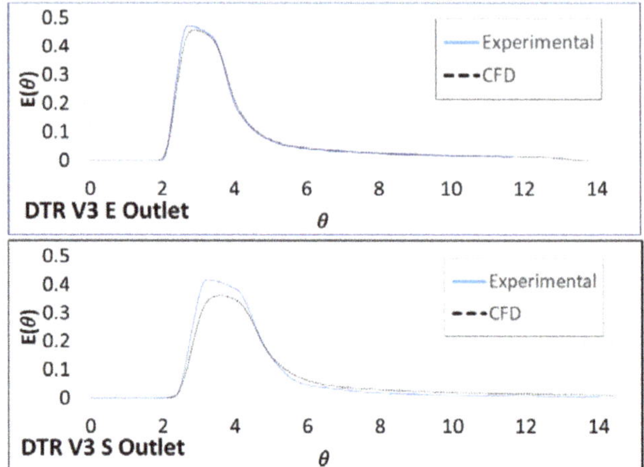

Figure 10. RTD curves for V3.

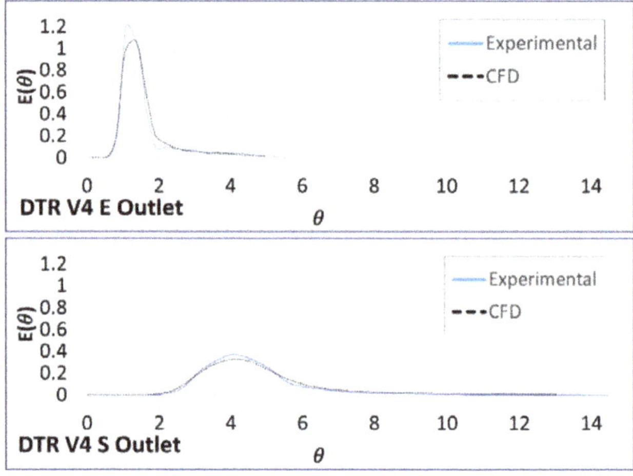

Figure 11. RTD curves for V4.

The variations between the experimental and theoretical RTD curves may be caused by the variability of the manual injection of the tracer, and tracer detection may be compared with the numerical convection-advection model on CFD. In almost all the cases, the peak of the experimental model is higher than in the CFD, resulting on the maximum concentration obtained in the experimental scenarios. Specially the Figure 11, where the RTD curve on the E outlet is too short when compared to the RTD curve on the S outlet, this is due to the outlet velocities that are registered on the experimental test, 1.25 m/s on the E outlet and 0.43 m/s on the S outlet, the tracer goes faster for the first velocity. The curves that were obtained for the validation tests V1 to V4 show that the tracer is following a tracer flow pattern that is similar to the experimental injection. To quantify these variations, an average error was calculated by the Root Mean Square Error (RMSE), Table 4.

Despite these differences, even the trend maintains a good proximity in terms of the distribution of the tracer, which shows that the proposed mathematical model is sufficiently calibrated model for describing the experimental mixture phenomenon in cross-junctions.

Table 4. Root Mean Square Error (RMSE) values for the four scenarios to evaluate variances in the RTD curves.

RMSE	V1	V2	V3	V4
E outlet	0.0215	0.0222	0.0093	0.0509
S outlet	0.0072	0.0109	0.0181	0.0140

3.4. Variation of Sc_T Coefficient

A variation of the turbulent diffusivity is presented, in the search to verify if the model, under these conditions of pressure and velocity, varies significantly if the Sc_T coefficient is modified. The turbulent diffusivity was simulated in different values, using assignments to the Schmidt Turbulent Sc_T number. The range of values for this parameter is 0.61, 0.71, 0.81, and the value that was obtained by the Kays-Crawford model [41], which amounts to 0.5666.

The results indicate that the model is not sensitive to the variation of these Sc_T coefficients, since the parameterization of the curves reflects practically homogeneous results. This conclusion is verified, if a RMSE for each case is obtained again (Table 5).

Table 5. Approximation of RTD curves varying Sc_T values.

RMSE	Sc_T (adim)			
	0.61	0.71	0.81	K-C (0.566)
V1 E	0.0215	0.0216	0.0216	**0.0215**
V1 S	0.0073	0.0074	0.0076	**0.0072**
V2 E	0.0223	0.0226	0.02282	**0.0220**
V2 S	0.0109	0.0107	0.0106	**0.0106**
V3 E	0.0090	0.0085	0.0081	**0.0093**
V3 S	0.0182	0.0184	0.0186	**0.0181**
V4 E	0.0509	0.0509	0.0509	**0.0509**
V4 S	0.0143	0.0147	0.0151	**0.0140**

Table 5 shows that the variation is verified of the value used of Sc_T was based on the model of Kays-Crawford [41], and it was the most appropriate in most cases. The values of Sc_T have a very low impact in scale for the simulated model. In Figure 12, it is observed that the diffusion increases in the zones of greater recirculation, and there is less speed in the model. This should have a more significant influence, according to the bibliography. For the moment, it can be determined that the speed and pressure range far exceed the studied cases, and this tells us that at high levels of speed and pressure, the turbulent diffusivity does not have a significant impact on the transport of solute.

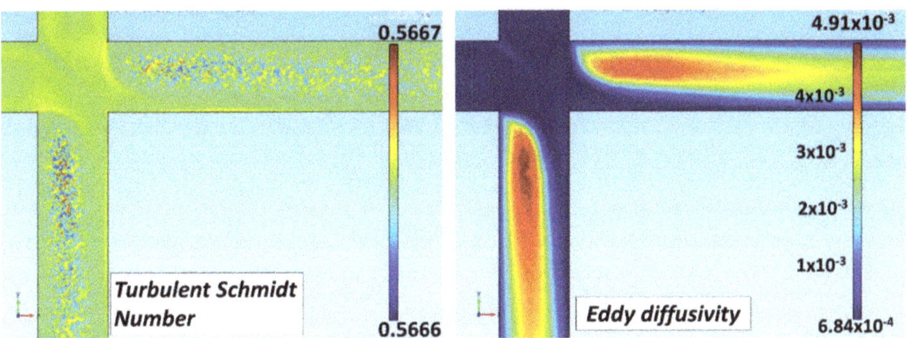

Figure 12. Sc_T and eddy diffusivity plots based on the Kays-Crawford [41] model (Equation (12)). High values appear in the turbulent flow zones.

3.5. Incomplete Mixing Simulations

Once the CFD model has been calibrated, the mixing grade can be evaluated for diverse scenarios with the tracer injection at the two inlets, N and W. To represent the development of the mixing graphically, it was introduced by two parameters: Inlet relation (C_N/C_W); in the same way, Outlet relation (C_E/C_S). The diverse scenarios implemented were made by the combination of concentration at the inlets, in intervals of 0.25 mol/L, in a range Inlet relation = [0.00–2.00]. In the case of Inlet relation = 0.00, it means that C_N = 0.00 mol/L and C_W = 5 mol/L. At the same way, for an Inlet relation = 1.50, it means that C_N = 7.5mol/L and C_W = 5mol/L. The scenarios are represented in Figure 13, with the results for the outlet relation after the cross-junction mixing.

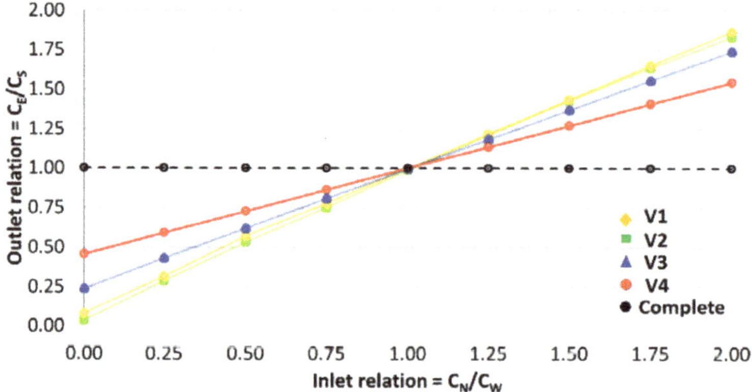

Figure 13. Solute mixing for variable concentration at the inlets.

The four curves V1–V4 that are shown in Figure 13 represent the concentration relation between the two outlets E and S. Each curve (apparently straight lines) has different slopes. It can be observed that curves V1 and V2 have a similar development pattern, and are different from the curves V3 and V4. This is because V1 and V2 have closer input and output flow velocities, around 1 m/s for each one. However, curves V3 and V4 correspond to the input and output velocities with a wide degree of difference, between 0.427 to 1.302 m/s. This indicates that the transport of solute is by convective flow. This difference between velocities causes the tracer to flow at a higher rate at one output than at the other one. Therefore, the greater the difference between the input and output velocities, the greater the change in the slope of the generated curve.

Due to the behavior of the velocities mentioned above, the mixing behavior is maintained in a different proportion to the outlets in practically all of the cases. The complete mixing is not carried out on any scenario. Even the scenario with the same concentration in both inlets (Inlet Relation = C_N/C_W = 1) might not be a complete mixing. That is, since the result contains the same concentration at the inlets, it does not mean that a total mixture of the two incoming flows has been made. The complete mixing model assumes that "Outlet relation" must be equal to one in all cases, by the reason that the estimation assigns the same concentration at the two outlets. No one scenario has the behavior like a complete mixing.

4. Discussion and Conclusions

The physical characteristics of the experimental network, the pumping equipment, and the measuring devices used, allowed for achieving higher conditions of turbulence in the experimental model. To analyze the mixing that occurs in the cross-junction under these turbulence conditions, four scenarios were generated for flow velocities between 0.43 and 1.53 m/s; and for pressures around the cross-junction between 1.56 and 1.96 bar, we considered four kinds of similar velocities for the inlets and outlets boundaries around the cross-junction. It is common for these conditions to occur at these variations in real drinking water distribution networks.

The RTD curves present a novel and an adequate approximation, having RMSE values that are between 0.0072 and 0.0509. It means that the average distance of separation in some points of the curves carries this value. These separations occurred for the most part in the peaks of the curves. However, in the ascending and descending curvatures, they present a very precise similarity. In most of these regions, slopes are well represented, in all validation tests.

The variation of Sc_T solved many questions about the role of turbulent diffusion and the impact that it generates in these operating conditions. The RTD curves showed a minimal change between them. The change is almost inappreciable graphically. Therefore, more checks were obtained using the RMSE error for each curve, and the variation is seen up to the fourth decimal place of precision of the RMSE. This allows for concluding that convection is the main transport in diluted species, and that the diffusion does not affect much in the simulation of the tracer course through the cross-junction. Even with this, it was found that the model that was proposed by Kays-Crawford [41] was the adequate parameter when presenting the minimum RMSE values with respect to the turbulent Schmidt variation.

The mesh selection and the geometric construction were adequate, because the values that were derived from flow and average velocity in a contour surface correspond approximately to those that were calculated with analytical formulas. Also, the simulation times did not require extensive periods, and convergence was achieved without the presence of broken curves in the graphics. The graphics could be processed with a reasonable degree of affinity in the surface of symmetry. The hydrodynamic approach retains a low degree of error with respect to experimental measurements. The errors that were obtained at the inlet boundaries, for the pressure was between 0.502% to 1.478% on the four scenarios and at the outlets boundaries, for the velocity, was between 0.133% and 1.720%. It not only depended on the simulation in CFD and the mesh, but also on the continuity adjustments that were made in the experimental tests. These adjusted tests ensure that the entry costs were equivalent to the costs of exit in the trials, which was also kept in the CFD environment since one of the principles of its coding is based on the continuity equation for three dimensions.

Finally, the mixing grade that was evaluated for the scenarios with tracer injection at the two inlets, varying nine combinations, shows that the mixing behavior is maintained in a different proportion to the outlets in practically all the cases. This is due to the behavior of the velocities that are mentioned above; the complete mixing is not carried out on any scenario and it was verified with the RTD analysis, where they show us on the graphics of the RTD curves how the tracer was obtained on the outlet boundaries at different times and peaks. It also could be verified on the scenarios with the same concentration at the inlets, obtain the same concentration, but it does not mean that a total mixture of the two incoming flows has been made, and that was validated with the same RTD curves. No one of the scenario has any behavior like complete mixing.

Author Contributions: For research articles with several authors, a short paragraph specifying their individual contributions must be provided. The following statements should be used "Conceptualization, D.H.-C., J.L.N. and J.M.R.; Methodology, D.H.-C., J.L.N., P.A.L.-J. and J.M.R. ; Software, J.L.N. and J.M.R.; Validation, D.H.-C., M.R., J.L.N., P.A.L.-J. and J.M.R.; Formal Analysis, D.H.-C., X.D.-G., J.L.N., P.A.L.-J. and J.M.R.; Investigation, D.H.-C. and J.M.R.; Resources, X.D.-G., J.L.N., P.A.L.-J. and J.M.R.; Data Curation, D.H.-C. and M.R.; Writing-Original Draft Preparation, , D.H.-C. and J.M.R.; Writing-Review & Editing, D.H.-C., M.R., X.D.-G., J.L.N., P.A.L.-J. and J.M.R.; Visualization, D.H.-C., X.D.-G. and J.M.R.; Supervision, J.L.N., P.A.L.-J. and J.M.R.; Project Administration, J.M.R.; Funding Acquisition, X.D.-G., J.L.N., P.A.L.-J. and J.M.R.", please turn to the CRediT taxonomy for the term explanation. Authorship must be limited to those who have contributed substantially to the work reported.

Acknowledgments: To CONACYT for the Master and Ph.D. scholarships (417824 and 703220) to D.H.-C. and the Ph.D. scholarship (294038) to M.R.; To Universidad de Guanajuato for the financial support of the project No. 100/2018 of J.L.N.; To Engineering Division, Campus Guanajuato and Geomatics and Hydraulics Engineering Department for the financial support of this project; and finally, to SEP-PRODEP and UG for the financial support to publish this paper.

Conflicts of Interest: The authors declare no conflict of interest.

References

1. Ahn, J.C.; Lee, S.W.; Choi, K.Y.; Koo, J.Y. Application of EPANET for the determination of chlorine dose and prediction of THMs in a water distribution system. *Sustain. Environ. Res.* **2012**, *22*, 31–38.
2. Shanks, C.M.; Sérodes, J.B.; Rodriguez, M.J. Spatio-temporal variability of non-regulated disinfection by-products within a drinking water distribution network. *Water Res.* **2013**, *47*, 3231–3243. [CrossRef] [PubMed]
3. Vasconcelos, J.J.; Rossman, L.A.; Grayman, W.M.; Boulos, P.F.; Clark, R.M. Kinetics of chlorine decay. *Am. Water Works Assoc.* **1997**, *89*, 54. [CrossRef]
4. Ozdemir, O.N.; Ucak, A. Simulation of chlorine decay in drinking-water distribution systems. *J. Environ. Eng.* **2002**, *128*, 31–39. [CrossRef]
5. Geldreich, E.E. *Microbial Quality of Water Supply in Distribution Systems*; CRC Press LLC: Boca Raton, FL, USA, 1996; ISBN 1-56670-194-5.
6. Knobelsdorf, M.J.; Mujeriego, S.R. Crecimiento bacteriano en las redes de distribución de agua potable: Una revisión bibliográfica. *Ing. del Agua* **1997**, *4*, 17–28. [CrossRef]
7. Alcocer-Yamanaka, V.H.; Tzatchkov, V.G.; Arreguín-Cortés, F.I. Modelo de calidad del agua en redes de distribución. *Ing. Hidraul. Mex.* **2004**, *19*, 77–88.
8. Rodríguez, M.J.; Rodríguez, G.; Serodes, J.; Sadiq, R. Subproductos de la desinfección del agua potable: Formación, aspectos sanitarios y reglamentación. *Interciencia* **2007**, *32*, 749–756.
9. Wang, W.; Ye, B.; Yang, L.; Li, Y.; Wang, Y. Risk assessment on disinfection by-products of drinking water of different water sources and disinfection processes. *Environ. Int.* **2007**, *33*, 219–225. [CrossRef] [PubMed]
10. Castro, P.; Neves, M. Chlorine decay in water distribution systems case study-lousada network. *Electron. J. Environ. Agric. Food Chem.* **2003**, *2*, 261–266.
11. Parks, S.L.I.; VanBriesen, J.M. Booster disinfection for response to contamination in a drinking water distribution system. *J. Water Res. Plan. Manag.* **2009**, *135*, 502–511. [CrossRef]
12. Tabesh, M.; Azadi, B.; Rouzbahani, A. Optimization of chlorine injection dosage in water distribution networks using a genetic algorithm. *J. Water Wastewater* **2011**, *22*, 2–11.
13. Kansal, M.L.; Dorji, T.; Chandniha, S.K.; Tyagi, A. Identification of optimal monitoring locations to detect accidental contaminations. In Proceedings of the World Environmental and Water Resources Congress 2012: Crossing Boundaries, Albuquerque, NM, USA, 20–24 May 2012; pp. 758–776.
14. Hernández, D.; Rodríguez, J.M.; Galván, X.D.; Medel, J.O.; Magaña, M.R.J. Optimal use of chlorine in water distribution networks based on specific locations of booster chlorination: Analyzing conditions in Mexico. *Water Sci. Technol. Water Supply* **2016**, *16*, 493–505. [CrossRef]
15. Weickgenannt, M.; Kapelan, Z.; Blokker, M.; Savic, D.A. Risk-based sensor placement for contaminant detection in water distribution systems. *J. Water Res. Plan. Manag.* **2010**, *136*, 629–636. [CrossRef]
16. Rathi, S.; Gupta, R. Monitoring stations in water distribution systems to detect contamination events. *ISH J. Hydraul. Eng.* **2014**, *20*, 142–150. [CrossRef]
17. Seth, A.; Klise, K.A.; Siirola, J.D.; Haxton, T.; Laird, C.D. Testing contamination source identification methods for water distribution networks. *J. Water Res. Plan. Manag.* **2016**, *142*. [CrossRef]
18. Xuesong, Y.; Jie, S.; Chengyu, H. Research on contaminant sources identification of uncertainty water demand using genetic algorithm. *Cluster Comput.* **2017**, *20*, 1007–1016. [CrossRef]
19. Montalvo, I.; Gutiérrez, J. Water quality sensor placement with a multiobjective approach. In Proceedings of the Congress on Numerical Methods in Engineering, Valencia, Spain, 3–5 July 2017.
20. Rathi, S.; Gupta, R. Optimal sensor locations for contamination detection in pressure-deficient water distribution networks using genetic algorithm. *Urban Water J.* **2017**, *14*, 160–172. [CrossRef]
21. Rossman, L.A. *EPANET 2: User's Manual*; EPA/600/R-00/057; U.S. Environmental Protection Agency: Washington, DC, USA, 2000.
22. Sandoval, M.; Rosalba, F.; Walsh, F.C.; Nava, J.L.; Ponce de León, C. Computational fluid dynamics simulations of single-phase flow in a filter-press flow reactor having a stack of three cells. *Electrochim. Acta* **2016**, *216*, 490–498. [CrossRef]
23. Castañeda, L.F.; Antaño, R.; Rivera, F.F.; Nava, J.L. Computational fluid dynamic simulations of single-phase flow in a spacer-filled channel of a filter-press electrolyzer. *Int. J. Electrochem. Sci.* **2017**, *12*, 7351–7364. [CrossRef]

24. Rodríguez, I.L.; Valdés, J.A.; Alfonso, E.; Estévez, R.D. Dopico. Uso de técnicas estímulo—Respuesta para simular diagnósticos en esófago humano. In Proceedings of the Congreso Latinoamericano de Ingeniería Biomédica, La Habana, Cuba, 23–25 May 2001.

25. Van Bloemen Waanders, B.; Hammond, G.; Shadid, J.; Collis, S.; Murray, R. A comparison of Navier-Stokes and network models to predict chemical transport in municipal water distribution systems. In Proceedings of the World Water and Environmental Resources Congress, Anchorage, AK, USA, 15–19 May 2005; pp. 1–10.

26. Webb, S.W. High-fidelity simulation of the influence of local geometry on mixing in crosses in water distribution systems. In Proceedings of the ASCE World Water & Environmental Resources Congress, Tampa, FL, USA, 15–19 May 2007.

27. Ho, C.K.; Khalsa, S.S. EPANET-BAM: Water quality modeling with incomplete mixing in pipe junctions. In Proceedings of the Water Distribution Systems Analysis 2008 Conference, Kruger National Park, South Africa, 17–20 August 2008; pp. 1–11.

28. Song, I.; Romero-Gomez, P.; Choi, C.Y. Experimental verification of incomplete solute mixing in a pressurized pipe network with multiple cross-junctions. *J. Hydraul. Eng.* **2009**, *135*, 1005–1011. [CrossRef]

29. Liu, H.; Yuan, Y.; Zhao, M.; Zheng, X.; Lu, J.; Zhao, H. Study of Mixing at Cross-junction in Water Distribution Systems Based on Computational Fluid Dynamics. In Proceedings of the International Conference on Pipelines and Trenchless Technology, Beijing, China, 26–29 October 2011; pp. 552–561.

30. Romero-Gomez, P.; Lansey, K.E.; Choi, C.Y. Impact of an incomplete solute mixing model on sensor network design. *J. Hydroinform.* **2011**, *13*, 642–651. [CrossRef]

31. Yu, T.C.; Shao, Y.; Shen, C. Mixing at cross joints with different pipe sizes in water distribution systems. *J. Water Res. Plan. Manag.* **2014**, *140*, 658–665. [CrossRef]

32. Shao, Y.; Yang, Y.J.; Jiang, L.; Yu, T.; Shen, C. Experimental testing and modeling analysis of solute mixing at water distribution pipe junctions. *Water Res.* **2014**, *56*, 133–147. [CrossRef] [PubMed]

33. Mompremier, R.; Pelletier, G.; Mariles, Ó.A.F.; Ghebremichael, K. Impact of incomplete mixing in the prediction of chlorine residuals in municipal water distribution systems. *J. Water Supply Res. Technol.* **2015**, *64*, 904–914. [CrossRef]

34. Mckenna, S.A.; O'hern, T.; Hartenberger, J. Detailed Investigation of Solute Mixing in Pipe Joints Through High Speed Photography. *Water Distrib. Syst. Anal.* **2008**, 1–12. [CrossRef]

35. Ho, C.K.; O'Rear, L., Jr. Evaluation of solute mixing in water distribution pipe junctions. *Am. Water Works Assoc. J.* **2009**, *101*, 116. [CrossRef]

36. Choi, C.Y.; Shen, J.Y.; Austin, R.G. Development of a comprehensive solute mixing model (AZRED) for double-tee, cross, and wye junctions. *Water Distrib. Syst. Anal.* **2008**, 1–10. [CrossRef]

37. Gualtieri, C.; Jiménez, P.L.; Rodríguez, J.M. A comparison among turbulence modelling approaches in the simulation of a square dead zone. In Proceedings of the 19th Canadian Hydrotechnical Conference, Vancouver, BC, Canada, 9–14 August 2009.

38. Versteeg, H.K.; Malalasekera, W. *An Introduction to Computational Fluid Dynamics: The Finite Volume Method*; Pearson Education: New York, NY, USA, 2007.

39. Rosales, M.; Pérez, T.; Nava, J.L. Computational fluid dynamic simulations of turbulent flow in a rotating cylinder electrode reactor in continuous mode of operation. *Electrochim. Acta* **2016**, *194*, 338–345. [CrossRef]

40. Hanjalic, K. Two-Dimensional Asymmetric Turbulent Flow in Ducts. Ph.D. Thesis, University of London, London, UK, 1970.

41. Kays, W.M.; Crawford, M.E.; Weigand, B. *Convective Heat and Mass Transfer*; McGraw-Hill Education: New York, NY, USA, 2005.

42. Martinez-Solano, F.J.; Iglesias-Rey, P.L.; Gualtieri, C.; López Jiménez, P.A. Modelling flow and concentration field in rectangular water tanks. In Proceedings of the International Environmental Modelling and Software Society (iEMSs), Ottawa, ON, Canada, 5–8 July 2010.

43. Moncho-Esteve, I.J.; Palau-Salvador, G.; López Jiménez, P.A. Numerical simulation of the hydrodynamics and turbulent mixing process in a drinking water storage tank. *J. Hydraul. Res.* **2005**, *2*, 207–217. [CrossRef]

44. Georgescu, A.M.; Georgescu, S.C.; Bernad, S.; Coşoiu, C.I. COMSOL Multiphysics versus Fluent: 2D numerical simulation of the stationary flow around a blade of the Achard turbine. In Proceedings of the 3rd Workshop on Vortex Dominated Flows, Timisoara, Romania, 1–2 June 2007.

45. COMSOL. *CFD Module User's Guide*; Version 4.3bCOMSOL Multiphysics; COMSOL, Inc.: Burlington, MA, USA, 2013.
46. Moukalled, F.; Mangani, L.; Darwish, M. *The Finite Volume Method in Computational Fluid Dynamics*; Springer: Berlin, Germany, 2016; Volume 113.

 © 2018 by the authors. Licensee MDPI, Basel, Switzerland. This article is an open access article distributed under the terms and conditions of the Creative Commons Attribution (CC BY) license (http://creativecommons.org/licenses/by/4.0/).

Article

Computational Determination of Air Valves Capacity Using CFD Techniques

**Salvador García-Todolí [1], Pedro L. Iglesias-Rey [1], Daniel Mora-Meliá [2,*],
F. Javier Martínez-Solano [1] and Vicente S. Fuertes-Miquel [1]**

[1] Departamento de Ingeniería Hidráulica y Medio Ambiente, Universitat Politècnica de Valencia,
   46022 Valencia, Spain; sgtodoli@upv.es (S.G.-T); piglesia@upv.es (P.L.I.-R.);
   jmsolano@upv.es (F.J.M.-S.); vfuertes@upv.es (V.S.F.-M.)

[2] Departamento de Ingeniería y Gestión de la Construcción, Facultad de Ingeniería, Universidad de Talca,
   3340000 Curicó, Chile

* Correspondence: damora@utalca.cl; Tel.: +56-951-265-044

Received: 7 September 2018; Accepted: 8 October 2018; Published: 12 October 2018

Abstract: The analysis of transient flow is necessary to design adequate protection systems that support the oscillations of pressure produced in the operation of motor elements and regulation. Air valves are generally used in pressurized water pipes to manage the air inside them. Under certain circumstances, they can be used as an indirect control mechanism of the hydraulic transient. Unfortunately, one of the major limitations is the reliability of information provided by manufacturers and vendors, which is why experimental trials are usually used to characterize such devices. The realization of these tests is not simple since they require an enormous volume of previously stored air to be used in such experiments. Additionally, the costs are expensive. Consequently, it is necessary to develop models that represent the behaviour of these devices. Although computational fluid dynamics (CFD) techniques cannot completely replace measurements, the amount of experimentation and the overall cost can be reduced significantly. This work approaches the characterization of air valves using CFD techniques, including some experimental tests to calibrate and validate the results. A mesh convergence analysis was made. The results show how the CFD models are an efficient alternative to represent the behavior of air valves during the entry and exit of air to the system, implying a better knowledge of the system to improve it.

Keywords: air valve; CFD; hydraulic characterization; entrapped air

---

1. Introduction

One of the main problems related to the operation and start-up of water distribution systems is the presence of air inside the pipes. There are many causes giving rise to the presence of air pockets: filling and emptying operations, temporary interruptions of water supply, vortexes in pumping feed tanks, air inlet in points with negative pressure, inflow in air valves during the negative pressure wave of a hydraulic transient, release of the dissolved air in the water, etc.

The presence and movement of air in water distribution pipes causes problems in most cases. Air pockets inside pipes can generate the following issues: reduction of pipe cross section, even blocks; the generation of an additional head loss, which increases energy consumption of pumping groups; decrease of pumps' performance; loss of filters' efficiency; noise and vibration problems; corrosion inside pipelines by the oxygen transported by the air; important errors in flow meters; etc. These problems originate an irregular system operation. However, the most important effect derived from the presence of air inside the pipes is the surge pressure caused by entrapped air pockets. These high pressures have their origin in the low inertia of the air and its high compressibility. The movement of water, with much more inertia, can compress the air pockets and generate unacceptable pressures in the system.

Air movement inside the pipes results from a combination of different effects like the drag of water in its movement, the natural movement of the air pockets within the water caused by the difference of density, and the resistance of the air both by frictional effects and the effects of surface tension. This movement is a complex problem. Martin [1] developed the first study about important pressure surges than the air pockets can generate inside the pipes. He pointed out that the slug flow regime is the only biphasic flow that can lead to significant pressure surges. Later, Liou [2] developed one of the first models of filling pipes with irregular profiles. Previously, most models had been developed for very simple geometries: horizontal, vertical or inclined with constant slope. A more general model than the one proposed by Liou, including the presence of air valves, was developed by Fuertes [3].

One of the current research lines on the presence of air pockets in hydraulic transients is the modelling of filling and emptying processes of pipelines. During the filling process the volume of air pockets reached extreme values. In these conditions, significant increases in pressure can be generated. Hence the need to study models that represent the filling of pipelines [4–6].

More recently, the problem of the presence of air in the pipeline has been extended to drainage and sewer systems. The increasing intensity of rainfall makes rain drainage systems exposed to high water floods. These large flow rates result in extremely rapid filling of large pipes, many of which are not designed to support important internal pressures [7–9].

The most characteristic device for controlling the air in pipes is the air valve [10]. This device can perform three basic functions. The first one is to release small amounts of air that accumulate at high points of the system during normal operation. The control of accumulated air in the highest points of the pipes prevents reducing the cross section and the potential reduction of transportation capacity of the system. A second function is related to system ventilation during the filling and emptying processes of the pipelines. These processes require that important air flows are admitted or expelled from the system and so the air valves are the most suitable devices to perform this function. Finally, the third function is related to the protection of facilities against transient phenomena.

The hydraulic transient is crucial in the manufacture, design and installation of pipelines, but to date, these effects have often been overlooked or poorly studied, mainly due to the difficulty of their evaluation (only through the use of complex mathematical models). The devices generally used in pressurized water pipes to manage the air inside the pipe are the air valves, because under certain circumstances, these can be used as an indirect control mechanism of a hydraulic transient. However, specific technical manuals do not abound and there is no firm policy in this regard to help engineers make decisions to select the most appropriate air valve. Consequently, it is necessary to develop tools to understand the behaviour of an air valve in an installation.

Most previous work on air valves in water facilities has focused on their ability to reduce the effects of hydraulic transients. In this regard, Stephenson [11] relates the correct selection of air valve size and standpipe used with water hammer minimization; Bianchi et al. [12] propose several simplified (experimentally validated) formulas for dimensioning the required air valve size during filling of a system; De Martino et al. [13] study the transient caused by the expulsion of air through an orifice and deduce a simple relationship to predict the maximum pressure during the transient which agrees well with experimental data; and Fuertes et al. [14] present a methodology for dimensioning air valves to control transient phenomena, but considering the strong transient effects caused by the closing of the air valve's float [13]. Finally, some researchers have studied the use of aeration devices in other applications of hydraulic engineering. In this regard, Bhosekar et al. [15] study the use of aerators in spillways in an application similar to [11], but in another field.

In short, the analysis of the literature does not allow see any previous study related with the use of the computational fluid dynamics (CFD) techniques for the characterization of the behavior of the air valves. Unfortunately, one of the major limitations is the reliability of information provided by manufacturers. This information can only be validated at the present through trials [16]. Since air valve tests are extremely expensive, the search for alternative techniques for characterizing these elements becomes a key point in this scientific field. Therefore, the goal of this work is to determine the capacity of CFD techniques to predict the behavior of commercial air valves. The work was divided into two parts: the numerical study made using a CFD code developed by ANSYS Fluent and the experimental study of the behavior of the air valves. In both cases, it is intended to obtain an experimentally determined relationship between the mass flow G and the pressure p_t inside the pipe where the air valve is connected. The final comparison of both studies (numerical and experimental) finally allows to see the validity of the proposed methodology as an alternative to the testing of this type of elements.

2. Characterization of Air Valves

The characteristic curve of an air valve is the relationship between the inhaled or exhaled mass flow G and the pressure p_t inside the pipe in the point where it is connected. There are different models to characterize the behaviour of an air valve, but they all assume that the air inside the air valve has compressible fluid behaviour. Regardless of the chosen model, it is necessary to determine certain parameters of the air valve, which vary according to the mathematical expression used for the characterization.

Traditionally, one of the most commonly used models resembles the behaviour of the fluid inside the air valve to the isentropic flow that occurs in a convergent nozzle (Figure 1). Wylie and Streeter [17] made the analogy between nozzles and air valves, where the air behaviour into the pipe is considered isothermal, while the airflow throughout the valve (both inlet and outlet flow) is assumed to be adiabatic.

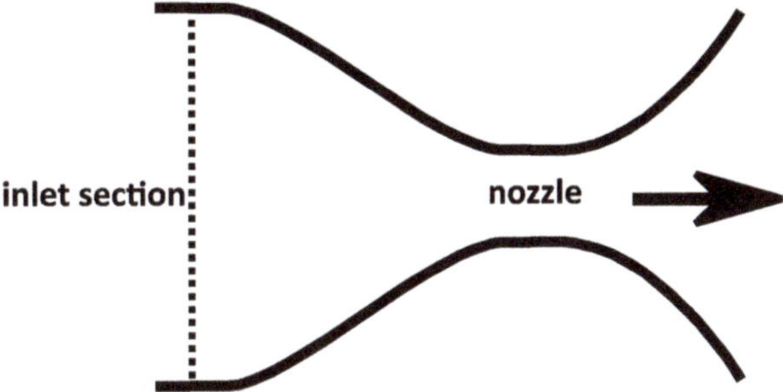

Figure 1. Isentropic flow in a convergent–divergent nozzle.

In this model, the air outlet mass flow G of an air valve, when the flow is subsonic is:

$$G = C_{exp} A_{exp} p_t^* \sqrt{\frac{7}{RT_t} \left[\left(\frac{p_{atm}^*}{p_t^*} \right)^{1.4286} - \left(\frac{p_{atm}^*}{p_t^*} \right)^{1.714} \right]} \tag{1}$$

In the Equation (1) A_{exp} is the air valve output cross section; p_t^* is the absolute pressure in the pipe upstream the air valve, representing the input pressure p_0 in Figure 1; R is the air characteristic constant in the classical thermodynamic formulation of an ideal gas; T_t is the air temperature inside the pipe; p_{atm}^* is the absolute value of the atmospheric pressure; and C_{exp} is the outlet characteristic coefficient of the air valve. This coefficient takes values less than one, and the lower the value the higher the airflow resistance.

When the air velocity is greater than the speed of sound (Mach number greater than 1), volumetric flow rate is blocked because speed cannot increase more. Thus, the mass flow rate can be higher since an increase in pressure causes an increase in the density of air. In these supersonic conditions, the mass flow G is:

$$G = C_{exp} A_{exp} \frac{0.686}{\sqrt{RT_t}} p_t^*$$

(2)

A similar approach can be carried out in the case of air inlet. In this case, there are two differences. On one side, the pressure in the inlet section is constant and equal to the atmospheric pressure. On the other side, the pressure in the output section is variable. This output section corresponds with the point of the pipe connected with the air valve. Therefore, the air inlet mass flow G, when the flow is subsonic is:

$$G = C_{adm} A_{adm} \sqrt{7 p_{atm}^* \rho_{atm} \left[\left(\frac{p_t^*}{p_{atm}^*} \right)^{1.4286} - \left(\frac{p_t^*}{p_{atm}^*} \right)^{1.714} \right]}$$

(3)

In Equation (3), C_{adm} is the inlet characteristic coefficient of the air valve, having the same considerations as C_{exp} in the output process; A_{adm} is air valve input cross section; and ρ_{atm} is the air density at atmospheric conditions.

If the flow is supersonic, the volumetric flow is kept constant. In this case, as the inlet pressure is constant (atmospheric pressure) the mass flow rate also remains constant. Thus, the flow is blocked. This means that even if the pressure inside the pipe decreases more, the amount of air admitted will not increase. Thus, in the conditions of supersonic flow G is:

$$G = C_{adm} A_{adm} \frac{0.686}{\sqrt{RT_{adm}}} p_{atm}^* = constant$$

(4)

Expressions (1)–(4) are theoretical formulations of the potential behaviour of an air valve. The real behaviour of the air valve must be obtained through tests, and this information should be provided by the manufacturer. As an example, Figure 2 shows a graph built from the technical information provided by a manufacturer.

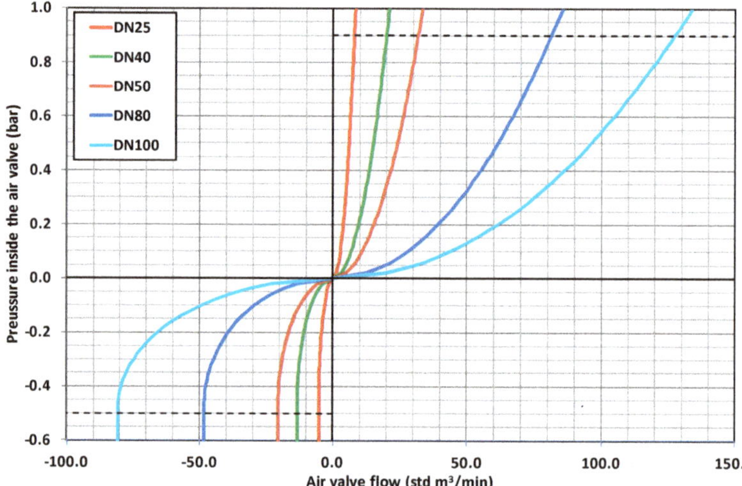

Figure 2. Air valves technical information provided a manufacturer.

Often, simplified expressions are used to treat numerically the characteristic curves of an air valve instead of the previous theoretical equations. The need to simplify these equations is related to decreasing the number of parameters on which the behaviour of the device depends. In this regard, Boldy [18] proposes a representation based on the equations of incompressible flow. Later, Fuertes [3] performed a comprehensive review of the different models representing the behaviour of an air valve. All simplified models seek to reduce the number of parameters needed to characterize this, setting a relationship between the mass flow admitted or expelled and the pressure inside the pipe.

Normally, the manufacturers of air valves present a graph of the characteristic curve. This is the ratio between the admitted/ejected airflow and the difference in pressure between the inside and outside of the air valve. This curve is obtained with experimental tests in all possible operating regions. It is possible to obtain some equations from it, which relate the airflow to the difference in pressures. Unfortunately, the conditions under which these tests are performed are often not referenced in the technical information provided by the manufacturers, so it is nearly impossible to reproduce these tests in a laboratory. In fact, previous studies [3] have already revealed the discrepancies between the commercial data provided by the manufacturer and the actual data obtained by testing. Figure 3 represents the differences between the experimental data obtained in the laboratory and the curve obtained from the technical information provided by the manufacturer.

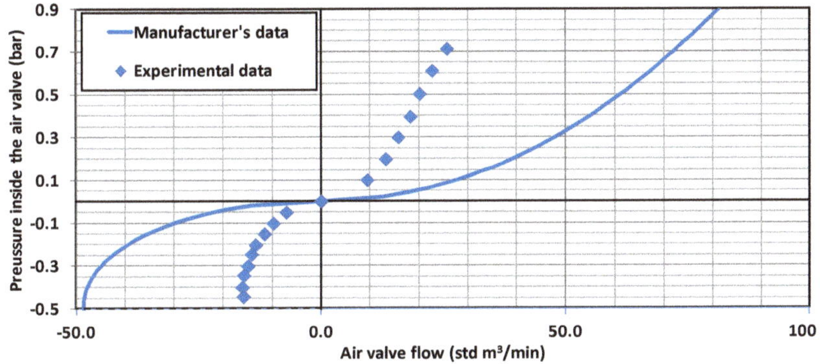

Figure 3. Comparison between laboratory tests and information provided by a manufacturer.

Results such as those in Figure 3 shows the need to find methodologies that allow characterizing the behaviour of air valves with sufficient reliability. This work proposes the use of CFD programs as an alternative, because these techniques are now considered as standard tools to predict the fluid flow behaviour. Calculations obtained using CFD will be compared with experimental tests performed in the laboratory.

The main problem of testing the characteristics of an air valve is the volume of air required. A 80 mm air valve can require a flow rate of 6200 m^3/h measured under standard conditions. If the size is 100 mm the required flow is 9700 m^3/h, while for 300 mm it may be necessary up to 87,000 m^3/h.

Currently there are two main techniques to test an air valve. The first is based on the storage of large quantities of air in high pressure tanks. During the test the air is released gradually through a system that reduces the pressure to the usual operating pressures of these elements. The difficulty of this system is the tank volumes required are very high (32 m^3 for 100 mm and almost 300 m^3 for a 300 mm). An alternative option is to use a blower capable of providing the flow and pressure necessary to perform the test. The problem with these installations is the cost associated with the construction of this type of equipment. Simply as a reference data, to test a 300 mm air valve capable of dealing 24 m/s with a pressure of 0.9 bar, requires an approximate power of 1.4 MW. This data is an indicator of the complexity of this type of facility and justifies the fact that there are few places to carry out this type of test.

The experimental study was conducted at the Air Valves Test Bench of Bermad (Israel). This installation has a blower of 315 kW with capacity of 16,000 m^3/h of air measured in standard conditions a 0.6 bar of differential pressure. At each test point the flow stability conditions were verified and subsequently the mass flow rate and the pressure at the inlet were recorded. The measuring devices have been previously calibrated guaranteeing errors below 0.5%. The system complies with all European standards and allows the testing of air valves between 2 and 12 inches in a continuous process that allows the measurement of air flow over the entire range of pressures considered. More details about the test bench and the collected data can be found in [16].

In this case, the tests were carried out on more than 20 different models from 13 different manufacturers, with flow ranges between -2500 and 3500 m^3/h and pressure ranges between -0.57 and 0.58 bar. All the elements tested were air/vacuum valves 80 mm diameter, capable of exhausting air during pipeline filling and admitting large amounts of air if pressure drops below atmospheric.

3. Computational Fluid Dynamics (CFD) Modelling

CFD means the use of a computer-based tool for simulating the behaviour of systems involving fluid flow, heat transfer, and other related physical processes. CFD works by solving the Navier–Stokes equations over a region of interest, describing how the different properties (velocity, pressure, temperature, density, etc.) of a moving fluid change. CFD is used by engineers and scientists in a wide range of fields, including motor industry (combustion modelling and aerodynamics), buildings (thermal comfort, fire effects, or and ventilation), electronics (heat transfer within and around circuit boards) or medical applications (cardiovascular medicine). In recent years, CFD models have been applied for solving several hydraulic engineering problems. In this way, CFD has been used for a variety of purposes, as simulations of flow transient in pipes [19,20], behaviour of different control valves under different conditions [21–24], mixing models in water distribution systems [25–28], or for the characterization of elements in open channels [29–34].

CFD allows the analysis of flow patterns that are difficult, expensive or impossible to study using experimental techniques and although they cannot completely replace the measurements, the amount of experimentation and the overall cost can be significantly reduced. In addition, the level of detail that can be achieved with these techniques is very high, generating a lot of additional information without added costs. This allows more complex and complete studies of those that can be performed experimentally.

CFD techniques also have a number of disadvantages. On the one hand, the results of a CFD simulation are not 100% reliable, since it is necessary to simplify real physical systems and properly model complex phenomena such as turbulence. On the other hand, the CFD techniques require computers with great calculation capacity. Since computer capabilities have increased in the last years, CFD techniques provide a tool for determining the pneumatic characteristics of an air valve.

To perform a CFD analysis, there are three main stages: pre-processing, simulation and post-processing. The pre-processing step includes geometry definition, mesh generation and boundary conditions definition. Once the physics problem has been identified, the simulation stage consists in solving the governing equations related to flow physics problems. Finally, the post-processing is the analysis of the results.

The geometry definition consists of specifying the shape of the physical boundaries of the fluid. Among other issues, it is necessary to define whether the computational model will be two-dimensional or three-dimensional. In this work, a standard computer-aided design (CAD) program is used to model a 3D commercial air valve. It should be noted that it is not necessary to include all the details of the air valve. Thus, given that the nominal diameter of the element is 80 mm, details of the geometry of size less than 1 mm have not been considered. Additionally, the geometry of the air valve (Figure 4) has been considered, so only half of it is represented. Likewise, in order to guarantee that the inflow into the air valve has been established, a straight pipe of the length equal to four diameters has been defined at the entrance.

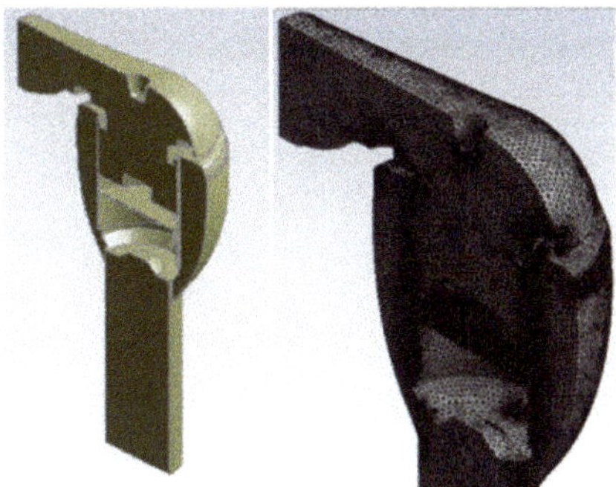

Figure 4. Geometry and meshing structure of a commercial air valve for ANSYS Fluent analysis.

The second step is mesh generation, which consists in dividing the domain of the fluid into a number of smaller cells. The solver used (ANSYS Fluent) is based on the finite volume method, where the domain is divided into a finite set of control volumes. The general conservation equations for mass and momentum are solved on this set of control volumes.

For the representation of the behaviour of the air valves, a structured tetrahedral mesh has been selected, since this type of mesh allows a better adjustment than the hexahedral meshes, especially in the curved areas of the interior of the valve. In order to ensure proper convergence of the model, a preliminary analysis of the minimum mesh size required was performed. This analysis is performed considering the following boundary conditions: the mass flow rate of air is fixed in the inlet while the pressure remains constant in the output. Specifically in this study the required pressure at the inlet of the valve to generate a flow of 2000 m^3/h of air under standard conditions, i.e., a mass flow of 0.667 kg/s, has been calculated.

All the simulations have been carried out by means of an implicit formulation in steady state. The resolution algorithm uses a density-based method, which has a coupled formulation of the equations of continuity, momentum and energy. This method of resolution allows better representation of the flow than pressure-based algorithms since the latter erroneously characterize negative pressure gradients. The numerical results of the convergence analysis of the mesh are collected in Table 1, where the value of the pressure P required at the inlet of the suction cup is shown based on the reference size used in the mesh and the number of elements it contains. As can be seen in Figure 5, mesh sizes around 4 mm are enough to analyze this type of problem with an adequate level of accuracy. Consequently, considering at least 300,000 cells guarantees an analysis with sufficient precision, as can be seen in Figure 6.

Table 1. Results of the mesh convergence analysis.

Size (mm)	No. Elements	P (bar)	Size (mm)	No. Elements	P (bar)	Size (mm)	No. Elements	P (bar)
10	21,026	0.295	6	93,806	0.341	3.8	363,719	0.34
9.5	24,462	0.308	5.5	120,654	0.334	3.6	428,428	0.339
9	28,227	0.303	5	160,731	0.343	3.4	507,391	0.34
8.5	33,851	0.311	4.8	181,723	0.344	3.2	608,079	0.34
8	40,060	0.309	4.6	205,557	0.342	3	738,175	0.34
7.5	48,426	0.315	4.4	234,307	0.339	2.8	907,558	0.34
7	59,363	0.32	4.2	270,010	0.339	2.5	1,272,477	0.34
6.5	73,917	0.332	4	311,827	0.339	2	2,481,359	0.34

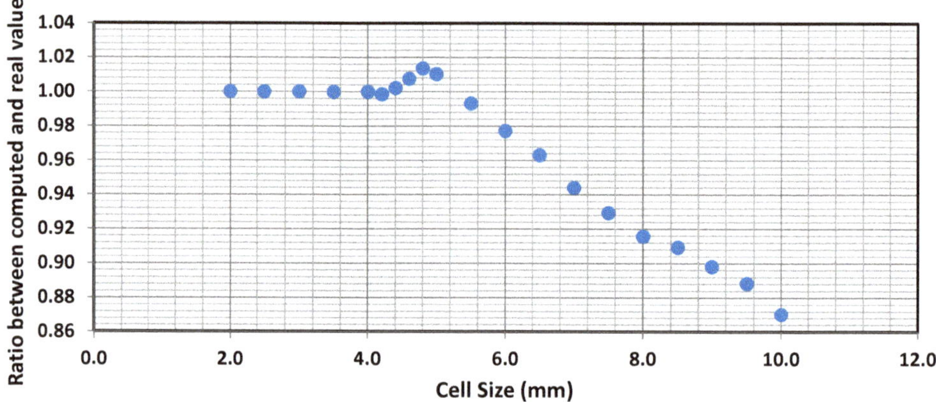

Figure 5. Analysis of the influence of the size of the mesh (size of the cells).

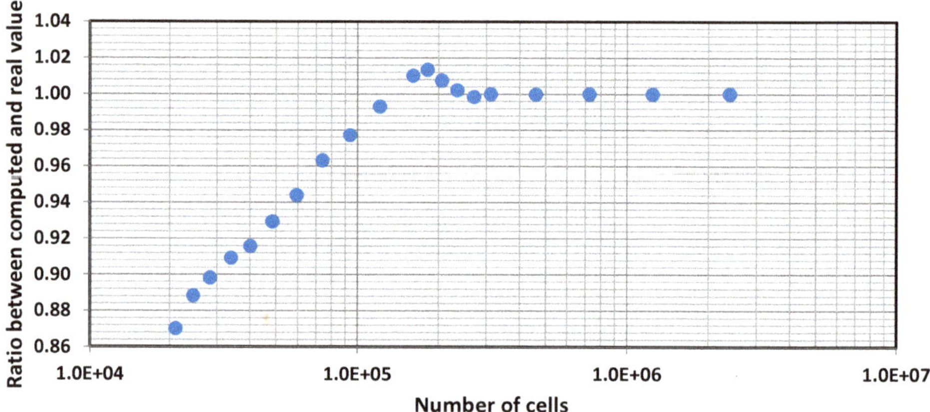

Figure 6. Analysis of the influence of the size of the mesh (number of cells).

Since a tetrahedral mesh has been used instead a hexahedral mesh, it has been necessary to review the aspect ratio of the different mesh elements. For the cell reference size finally selected (4 mm), 95% of the elements have a ratio of less than 2.5. Although only 5% of the cells have higher values, considering that less than 0.5% of the elements have an aspect ratio greater than 4.

The equations that describe the properties and motion of the fluid are numerically solved in each of the defined mesh cells. In general, these equations are obtained by applying the equations of continuity and momentum. The application of these laws allows obtaining the Navier–Stokes equations, which for a compressible flow, such as the air inside an air valve, can be expressed in a simplified way such as:

$$\frac{\partial \rho}{\partial t} + \nabla \cdot \left(\rho \vec{v} \right) = 0 \tag{5}$$

$$\rho \frac{\partial \vec{v}}{\partial t} + \rho \left(\vec{v} \nabla \right) \vec{v} = -\nabla p + \rho \vec{g} + \nabla \cdot \tau_{ij} \tag{6}$$

where ρ is the density of the fluid, $\vec{v}$ is the velocity field, p is the pressure field, g are the forces per unit volume to which the fluid is subjected and τ_{ij} is the stress tensor.

The Navier–Stokes equations are analytical and in order to solve them with a computer it is necessary to convert them into an algebraic system. This process is known as numerical discretization and there are different methods, being the most used finite difference, finite elements and finite volume.

Another of the basic aspects to consider when performing CFD analysis is the definition of the turbulence model. The objective of the turbulence models for the Reynolds-averaged Navier–Stokes (RANS) equations is to compute the Reynolds stresses and this is one of the basic aspects to consider when analyzing a flow through CFD.

There are several different formulations for solving turbulent flow problems such as Spalart–Allmaras, *k*-epsilon, *k*-omega, Menter's Shear Stress Transport models, etc. All of these models increase the Navier-Stokes equations with an additional viscosity term. The differences between them are the use of wall functions, the number of additional variables solved and what these variables represent.

This work uses a *k*-ε two-equation turbulence model introduced by Launder and Spalding [35], where *k* is the turbulent kinetic energy and ε is the rate of dissipation of kinetic energy. The *k*-ε model has great applicability to turbulent air and water flows and is very popular for practical engineering applications, due to its good convergence and relatively less computational requirements, obtaining results with comparable accuracy to higher order models.

Actually, the *k*-epsilon model would be more properly called a family of models, because some variants have been developed for so many specific flow configurations. In this work, a feasible *k*-ε realizable model has been used, where the term realizable means that the model satisfies certain mathematical constraints on the Reynolds stresses, consistent with the physics of turbulent flows. The modelled transport equations to find the terms *k* and ε are:

$$\frac{\partial}{\partial t}(\rho k) + \frac{\partial}{\partial x_j}(\rho k u_j) = \frac{\partial}{\partial x_j}\left[\left(\mu + \frac{\mu_t}{\sigma_k}\right)\frac{\partial k}{\partial x_j}\right] + G_k + G_b - \rho\varepsilon - Y_M + S_k \tag{7}$$

$$\frac{\partial}{\partial t}(\rho\varepsilon) + \frac{\partial}{\partial x_j}(\rho\varepsilon u_j) = \frac{\partial}{\partial x_j}\left[\left(\mu + \frac{\mu_t}{\sigma_\varepsilon}\right)\frac{\partial\varepsilon}{\partial x_j}\right] + \rho C_1 S_\varepsilon - \rho C_2 \frac{\varepsilon^2}{k + \sqrt{v\varepsilon}} + C_{1\varepsilon}\frac{\varepsilon}{k}C_{3\varepsilon}G_b + S_\varepsilon \tag{8}$$

In Equations (7) and (8) u_j represents velocity component in corresponding direction (x_j); μ_t is the absolute dynamic viscosity of the air; G_k is the turbulent kinetic energy generated by the variations of the mean flow velocity components; G_b represents the kinetic energy generated by the buoyancy effects; Y_M is the contribution of the pulsatile expansion associated with the compressible turbulence; $C_{1\varepsilon}$, C_2 and $C_{3\varepsilon}$ are constants; σ_k and σ_ε are the Prandl numbers for *k* and ε respectively; and S_k and S_ε represent a global variation in time of the parameters *k* and ε, being defined independently of the rest of variables. Likewise, the constants C_1, η and S are defined by:

$$C_1 = max\left[0.43, \frac{\eta}{\eta + 5}\right] \tag{9}$$

$$\eta = S\frac{k}{\varepsilon} \tag{10}$$

$$S = \sqrt{2S_{ij}S_{ij}} \tag{11}$$

On the other hand, turbulent viscosity μ_t is calculated by a combination of *k* and ε by the equation:

$$\mu_t = \rho C_\mu \frac{k^2}{\varepsilon} \tag{12}$$

where C_μ is a constant.

In this model, the constants $C_{1\varepsilon}$, C_2, $C_{3\varepsilon}$, σ_k and σ_ε are adopted from the values proposed by Launder and Spalding [35], since this proved that they are very effective for turbulent flows with a wide range of Reynolds number for both water and air. The realizable model has shown some improvements over the standard k-ε model, because it predicts more accurately the spreading rate of both planar and round jets. Likewise, it has also been shown that it performs better in flows involving rotation, boundary layers under strong adverse pressure gradients, separation and recirculation.

For a complete treatment of this problem, properties, initial and boundary conditions of the flow in space and time, would need to be specified. Regarding the properties of the fluid, it is necessary to specify viscosity, density and thermal properties. Regarding the flow inside the air valve, the viscosity of the air is considered constant, while the density varies admitting an air behaviour as if it were a perfect gas.

As initial conditions initial values for the variables are considered, from which the iterative process begins. The closeness of these initial values to the final solution has an influence on the process convergence time. In this work the initial pressure value is equal to the atmospheric pressure, while the initial air velocity is zero at all points.

The boundary conditions control the value of the variables or their relationship in the domain limits analyzed. It is basically about setting fixed values of pressure, velocity and temperature. On the model of an air valve of this work, the pressure is kept constant at the entrance and the exit. To analyze the air expulsion, this fixed inlet pressure will change in the different simulations, while the pressure in the outlet section will be constant and equal to the atmospheric pressure.

Once the model input values are completed, the software can solve the equations for each cell until an acceptable convergence is achieved.

The convergence of the method is based on the analysis of different criteria. Thus, as a general rule, the error is required to be less than 10^{-3} in the continuity equation, in all the Navier–Stokes momentum equations for each direction (x, y, z), and in the model-specific equations of turbulence (k and ε in this case). Additionally, given the compressible nature of the fluid, the use of the energy equation is required. In this case, the degree of convergence is more demanding (10^{-6}) than in the rest of convergence equations. Likewise, the fulfillment of a certain value of the residuals of the equations is not enough. Therefore, in order to ensure the convergence of the method, stability is required in the values of both the pressure at the inlet of the air valve and the mass flow that circulates inside the valve.

Finally, a post-processing software included in ANSYS is used to analyse the results generated by the CFD analysis. The data obtained can be analysed both numerically and graphically. In general, the most interesting graphic results that can be obtained is the distribution of pressures inside the air valve (Figure 7).

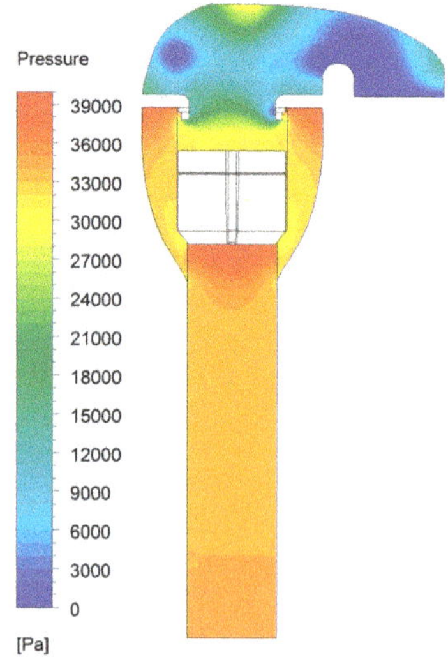

Figure 7. Pressure distribution in the midplane of the air valve.

Figure 7 allows analyzing in detail the different parts of the interior of the air valve where greater energy is lost. From the point of view of the design of the device, these zones are enlisted in areas where the pressure gradients are greater. Undoubtedly, improving the capacity of admission or expulsion of an air valve involves improving the design of these points. However, the objective of this work is not the design of this type of device, but to determine the capacity of the CFD techniques to predict their capacity of admission and expulsion.

4. Results

The application of the described analytical methodology allows comparing results obtained computationally with those obtained experimentally. For the particular case of one of the air valves studied, the results considering air as expulsion flow are shown in Figures 8 and 9.

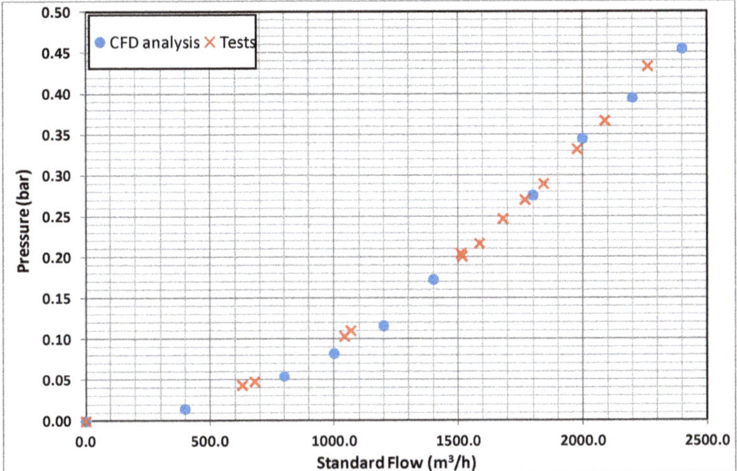

Figure 8. Comparison of the curve of an air valve: experimental data vs. computational fluid dynamics (CFD) analysis.

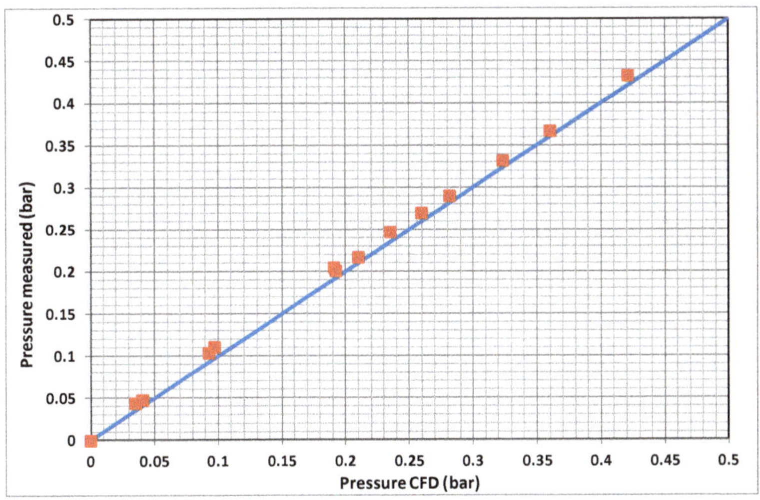

Figure 9. Comparison of the curve of an air valve: experimental data vs. CFD analysis.

The Figure 8 considers the representation of pressure and flow values. This representation demonstrates the good correspondence between the results obtained through the CFD simulation and the results obtained through laboratory tests. Additionally, Figure 9 represents the pressure values of the CFD results against the laboratory tests. This allows comparing both methodologies in a more detailed way. The results demonstrate the validity of the methodology based on CFD to be able to predict the behavior of an air valve.

Even considering the goodness of the results in Figure 9, to validate the CFD-based methodology of this work as an adequate technique for representing air inlet and outlet characteristics in an air valve, two additional questions need to be answered. On the one hand, it is necessary to validate the methodology for other geometries. On the other hand, it is also required to validate the method when it is applied in the case of air intake from the outside to the inside of the pipeline.

Related to this, the CFD analysis was performed on six different models of air valves of the same nominal diameter (DN 80 mm) but with different internal configurations and different air outlet mechanisms (down, side, mushroom). The results of these analyses are presented in Figure 10.

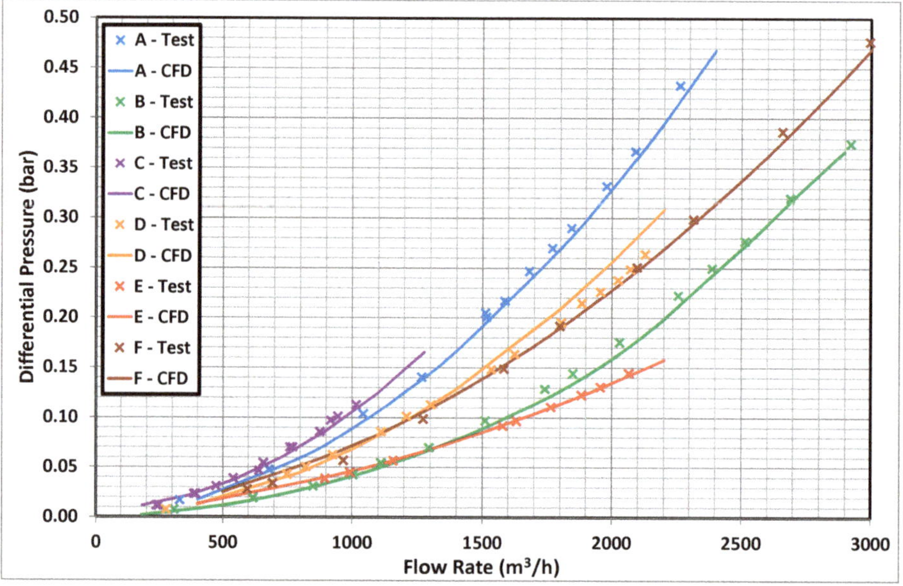

Figure 10. Comparison between laboratory tests and CFD simulations of six different models of air valve. Air expulsion flow.

The results show a good fit between CFD simulations and laboratory tests. However, all tests are performed up to a certain maximum pressure value. This maximum pressure is conditioned in each case for different reasons. In some cases, some of the air valves did not allow further extension of the air expulsion tests, since the flow that causes the dynamic closing of the valve was reached. That is to say, the air valve closes without the presence of water simply by the efforts caused by the air current that overcame the weight of the float. In other cases, the analysis is not extended because the maximum capacity of the test equipment is reached, which is 3000 m³/h of free air or 0.5 bar of differential pressure.

Once the capacity of the CFD model to represent the behavior of the air valve in ejection is verified, the process is repeated in the case of air intake. It should be noted that the compressibility effects of air are more noticeable when the pressures are lower. In this sense, it is important to highlight the capacity of the CFD model to represent the sonic block that is reached in the air valve when the pressures approach values close to −0.5 bar.

Figure 11 shows the results of the comparative study of six models of air valves in the intake phase. As can be observed, there are hardly any differences between the results obtained by laboratory tests and by CFD techniques.

As in the study of the expulsion of air, there is an analysis limit of 0.5 bar differential pressure between the outside of the pipeline and the inside of the pipeline. There are two reasons for this limitation of the study. The first is the maximum test capacity of the test bench used. The second is that for differential pressures greater than 0.5 bar below atmospheric pressure, the so-called sonic block is produced. In other words, the volumetric flow admitted by the suction cup reaches its maximum value. Consequently, as shown in Figure 11, both the results of the CFD model and the laboratory tests tend asymptotically to a value of maximum intake flow and this value will not increase although the depression inside the conduction increases.

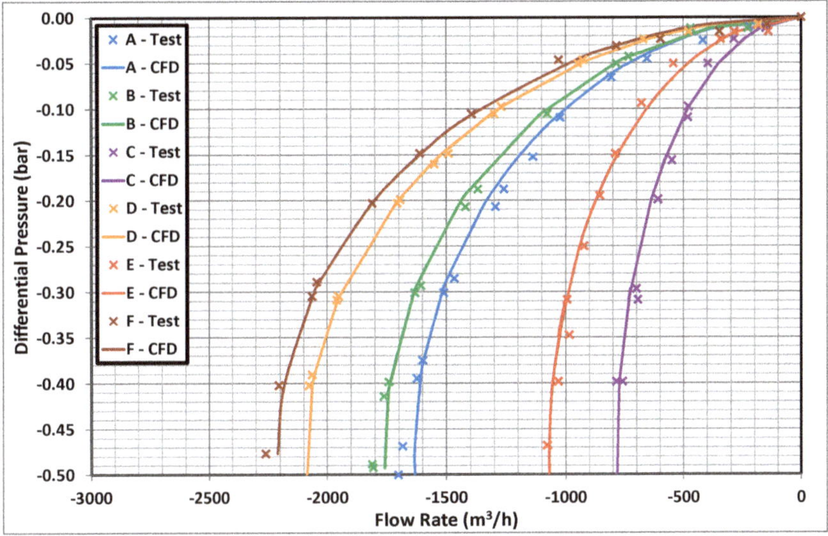

Figure 11. Comparison between laboratory tests and CFD simulations of six different models of air valve. Air intake flow.

5. Conclusions

CFD techniques provide a tool for determining in detail the behavioral characteristics of hydraulic elements. However, CFD requires a proper calibration of the models, so it will be essential to validate the results by tests in the laboratory. Specifically, this work shows the effectiveness of these techniques to represent the behavior of air valves that are installed in water supply networks. According to the results, it is possible to state the following:

- The size and quality of the mesh play a key role in the stability and accuracy of numerical calculations. The computational results of this work in 80 mm air valves reproduce accurately the laboratory tests with meshes from 250,000 elements, which is equivalent to a reference size equal to or less than 4 mm for each cell. The lower quality meshes introduce significant errors in the analysis, even higher than 10%.

- The CFD model has successfully represented the processes of air expulsion in an air valve, obtaining practically the same results in computer simulations as in laboratory tests. However, the CFD analysis does not allow prediction of the dynamic closure of the air valve. This dynamic closure has only been validated through laboratory tests. This is the reason why in some cases the curves of the air valves obtained through CFD cover a range of flow greater than those obtained experimentally.

- CFD models have been very effective representing the behavior of air valves during intake processes, where compressibility phenomena are more important. The CFD techniques can predict the behavior of the air valves in admission and specifically represent the sonic block when depressions close to 0.5 bar are reached. Under these circumstances, both the tests and the CFD simulations indicate the presence of a maximum flow rate that the air valve can accept.

- The discrepancies between the CFD model and the laboratory tests are lower than the error levels of the measuring devices used, in practically all the cases analyzed. Therefore, the correlation between the two techniques can be considered good. This undoubtedly highlights the adequacy of the type and size of mesh selected and the selected turbulence model.

To summarize, it can be concluded that the CFD models are an efficient alternative to represent the air valves during the entry and exit of air to the system. Given the difficulty and the costs of testing this type of element, the application of these techniques is an adequate alternative for the characterization of these elements. In the same way, the methodology could be very useful to verify the technical information provided by the manufacturers.

Author Contributions: All authors contributed extensively to the work presented in this paper. S.G.-T., P.L.I.-R. and D.M.-M., contributed to the subject of research, the modelling, the data analysis, and the writing of the paper. F.J.M.-S. and V.S.F.-M. contributed to the experimental measures and the manuscript review.

Funding: This research was funded by the Program Fondecyt Regular, grant number 1180660.

Acknowledgments: This work was supported by the Program Fondecyt Regular (Project 1180660) of the Comision Nacional de Investigación Científica y Tecnológica (Conicyt), Chile.

Conflicts of Interest: The authors declare no conflict of interest.

Abbreviations

The following abbreviations are used in this manuscript:

CFD   Computational Fluid Dynamics
RANS   Reynolds-averaged Navier–Stokes

References

1. Martin, C.S. Entrapped air in pipelines. In *Proceedings of the 2nd International Conference on Pressure Surges*; BHRA Fluid Engineering: London, UK, 1976; Volume 2, pp. 15–27.
2. Liou, C.P.; Hunt, W.A. Filling of Pipelines with Undulating Elevation Profiles. *J. Hydraul. Eng.* **1996**, *122*, 534–539. [CrossRef]
3. Fuertes, V.S. Hydraulic Transients with Entrapped Air Pockets. Ph.D. Thesis, Department of Hydraulic Engineering, Polytechnic University of Valencia, Valencia, Spain, 2001.
4. Zhou, F.; Hicks, F.E.; Steffler, P.M. Transient Flow in a Rapidly Filling Horizontal Pipe Containing Trapped Air. *J. Hydraul. Eng.* **2002**, *128*, 625–634. [CrossRef]
5. Laanearu, J.; Annus, I.; Koppel, T.; Bergant, A.; Vučković, S.; Hou, Q.; Tijsseling, A.S.; Anderson, A.; van't Westende, J.M.C. Emptying of Large-Scale Pipeline by Pressurized Air. *J. Hydraul. Eng.* **2012**, *138*, 1090–1100. [CrossRef]
6. Apollonio, C.; Balacco, G.; Fontana, N.; Giugni, M.; Marini, G.; Piccinni, A. Hydraulic Transients Caused by Air Expulsion During Rapid Filling of Undulating Pipelines. *Water* **2016**, *8*, 25. [CrossRef]
7. Zhou, F.; Hicks, F.E.; Steffler, P.M. Observations of Air–Water Interaction in a Rapidly Filling Horizontal Pipe. *J. Hydraul. Eng.* **2002**, *128*, 635–639. [CrossRef]
8. Vasconcelos, J.G.; Wright, S.J.; Roe, P.L. Improved Simulation of Flow Regime Transition in Sewers: Two-Component Pressure Approach. *J. Hydraul. Eng.* **2006**, *132*, 553–562. [CrossRef]

9.   Li, J.; McCorquodale, A. Modeling Mixed Flow in Storm Sewers. *J. Hydraul. Eng.* **1999**, *125*, 1170–1180. [CrossRef]

10.   Ramezani, L.; Karney, B.; Malekpour, A. The Challenge of Air Valves: A Selective Critical Literature Review. *J. Water Resour. Plan. Manag.* **2015**, *141*, 04015017. [CrossRef]

11.   Stephenson, D. Effects of Air Valves and Pipework on Water Hammer Pressures. *J. Transp. Eng.* **1997**, *123*, 101–106. [CrossRef]

12.   Bianchi, A.; Mambretti, S.; Pianta, P. Practical Formulas for the Dimensioning of Air Valves. *J. Hydraul. Eng.* **2007**, *133*, 1177–1180. [CrossRef]

13.   De Martino, G.; Fontana, N.; Giugni, M. Transient Flow Caused by Air Expulsion through an Orifice. *J. Hydraul. Eng.* **2008**, *134*, 1395–1399. [CrossRef]

14.   Fuertes-Miquel, V.S.; Iglesias-Rey, P.L.; Lopez-Jimenez, P.A.; Mora-Melia, D. Air valves sizing and hydraulic transients in pipes due to air release flow. In Proceedings of the 33rd IAHR Congress: Water Engineering for a Sustainable Environment, Vancouver, BC, Canada, 9–14 August 2009; IAHR: Vancouver, BC, Canada, 2009.

15.   Bhosekar, V.V.; Jothiprakash, V.; Deolalikar, P.B. Orifice Spillway Aerator: Hydraulic Design. *J. Hydraul. Eng.* **2012**, *138*, 563–572. [CrossRef]

16.   Iglesias-Rey, P.L.; Fuertes-Miquel, V.S.; García-Mares, F.J.; Martínez-Solano, J.J. Comparative Study of Intake and Exhaust Air Flows of Different Commercial Air Valves. *Procedia Eng.* **2014**, *89*, 1412–1419. [CrossRef]

17.   Wylie, E.B.; Streeter, V.L. *Fluid Transients in Systems*; Prentice Hall: Englewood Cliffs, NJ, USA, 1993.

18.   Boldy, A.P. The representation and use of air inlet/outlet valves for pressure surge control. In *Unsteady Flow Fluid Transients*; Watts, J., Bettess, R., Eds.; Balkema: Rotterdam, The Netherland, 1992.

19.   Martins, N.M.C.; Soares, A.K.; Ramos, H.M.; Covas, D.I.C. CFD modeling of transient flow in pressurized pipes. *Comput. Fluids* **2016**, *126*, 129–140. [CrossRef]

20.   Zhou, L.; Liu, D.; Ou, C. Simulation of Flow Transients in a Water Filling Pipe Containing Entrapped Air Pocket with VOF Model. *Eng. Appl. Comput. Fluid Mech.* **2011**, *5*, 127–140. [CrossRef]

21.   Davis, J.A.; Stewart, M. Predicting Globe Control Valve Performance—Part I: CFD Modeling. *J. Fluids Eng.* **2002**, *124*, 772. [CrossRef]

22.   Stephens, D.; Johnson, M.C.; Sharp, Z.B. Design Considerations for Fixed-Cone Valve with Baffled Hood. *J. Hydraul. Eng.* **2012**, *138*, 204–209. [CrossRef]

23.   Palau-Salvador, G.; González-Altozano, P.; Balbastre-Peralta, I.; Arviza-Valverde, J. Improvement in a Control Valve Geometry by CFD Techniques. In Proceedings of the ASCE Pipeline Division Specialty Conference, Houston, TX, USA, 21–24 August 2005; American Society of Civil Engineers: Reston, VA, USA, 2005; pp. 202–215.

24.   Lin, F.; Schohl, G.A. CFD Prediction and Validation of Butterfly Valve Hydrodynamic Forces. In *Critical Transitions in Water and Environmental Resources Management*; American Society of Civil Engineers: Reston, VA, USA, 2004; pp. 1–8.

25.   Romero-Gomez, P.; Ho, C.K.; Choi, C.Y. Mixing at Cross Junctions in Water Distribution Systems. I: Numerical Study. *J. Water Resour. Plan. Manag.* **2008**, *134*, 285–294. [CrossRef]

26.   Austin, R.G.; van Bloemen Waanders, B.; McKenna, S.; Choi, C.Y. Mixing at Cross Junctions in Water Distribution Systems. II: Experimental Study. *J. Water Resour. Plan. Manag.* **2008**, *134*, 295–302. [CrossRef]

27.   Liu, H.; Yuan, Y.; Zhao, M.; Zheng, X.; Lu, J.; Zhao, H. Study of Mixing at Cross Junction in Water Distribution Systems Based on Computational Fluid Dynamics. In Proceedings of the International Conference on Pipelines and Trenchless Technology (ICPTT), Beijing, China, 26–29 October 2011; American Society of Civil Engineers: Reston, VA, USA, 2011; pp. 552–561.

28.   Ho, C.K. Solute Mixing Models for Water-Distribution Pipe Networks. *J. Hydraul. Eng.* **2008**, *134*, 1236–1244. [CrossRef]

29.   Huang, J.; Weber, L.J.; Lai, Y.G. Three-Dimensional Numerical Study of Flows in Open-Channel Junctions. *J. Hydraul. Eng.* **2002**, *128*, 268–280. [CrossRef]

30.   Weber, L.J.; Schumate, E.D.; Mawer, N. Experiments on Flow at a 90° Open-Channel Junction. *J. Hydraul. Eng.* **2001**, *127*, 340–350. [CrossRef]

31.   Chanel, P.G.; Doering, J.C. Assessment of spillway modeling using computational fluid dynamics. *Can. J. Civ. Eng.* **2008**, *35*, 1481–1485. [CrossRef]

32.   Li, S.; Cain, S.; Wosnik, M.; Miller, C.; Kocahan, H.; Wyckoff, R. Numerical Modeling of Probable Maximum Flood Flowing through a System of Spillways. *J. Hydraul. Eng.* **2011**, *137*, 66–74. [CrossRef]

33. Castillo, L.; García, J.; Carrillo, J. Influence of Rack Slope and Approaching Conditions in Bottom Intake Systems. *Water* **2017**, *9*, 65. [CrossRef]
34. Regueiro-Picallo, M.; Naves, J.; Anta, J.; Puertas, J.; Suárez, J. Experimental and Numerical Analysis of Egg-Shaped Sewer Pipes Flow Performance. *Water* **2016**, *8*, 587. [CrossRef]
35. Launder, B.E.; Spalding, D.B. *Mathematical Models of Turbulence*; Academic Press: New York, NY, USA, 1972.

 © 2018 by the authors. Licensee MDPI, Basel, Switzerland. This article is an open access article distributed under the terms and conditions of the Creative Commons Attribution (CC BY) license (http://creativecommons.org/licenses/by/4.0/).

Case Report

Approaches to Failure Risk Analysis of the Water Distribution Network with Regard to the Safety of Consumers

Katarzyna Pietrucha-Urbanik * and Barbara Tchórzewska-Cieślak

Department of Water Supply and Sewerage Systems, Faculty of Civil, Environmental Engineering and Architecture, Rzeszow University of Technology, Al. Powstancow Warszawy 6, 35-959 Rzeszow, Poland; cbarbara@prz.edu.pl
* Correspondence: kpiet@prz.edu.pl; Tel.: +48-17-865-1703

Received: 22 September 2018; Accepted: 14 November 2018; Published: 17 November 2018

Abstract: Contemporary risk assessment makes reference to current world trends, whereby there is increased emphasis on safety. This paper has thus sought mainly to present new approaches to failure risk assessment where the functioning of a water distribution network (WDN) is concerned. The framework for the research involved here has comprised of: (a) an analysis of WDN failure in regard to an urban agglomeration in south-east Poland; (b) failure rate analysis, taking account of the type of a water pipe (mains, distribution, service connections (SC)) and months of the year, with an assessment of results in terms of criterion values for failure rate; (c) a determination—by reference to analyses performed previously—of the compatibility of experts' assessments in terms of standards of failure and obtained results, through rank analysis; and (d) the proposing of a modified Multi-Criteria Decision Analysis with implementation of an Analytical Hierarchy Process, to allow failure risk assessment for the WDN to be performed, on the basis of the calculated additive value of obtained risk. The analysis in question was based on real operating data, as collected from the water distribution company. It deals with failures in the WDN over a period of 13 years in operation, from 2005 to 2017.

Keywords: safety of water supply consumers; risk; water supply system; failure risk analysis; decision making model; risk assessment methodology

1. Introduction

The second half of the twentieth century brought many major accidents and disasters relating to the functioning of public water supply systems (WSS) in urban and industrial agglomerations. In this regard, there can be no doubt that WSS operations are subject to risk [1–6], hence the key importance of risk analysis that determines the location and size of such risk, as well as the actions to be taken with a view to its reduction or elimination [7]. Activities allowing these relations to be studied fall under the heading of risk management, which should be organized and comprehensive, in relation to both the entire WSS and elements associated with it [8]. The risk function is a measure of security loss [9]. There is always the possibility of a domino effect, so events are related to risk escalation [7]. In this regard, the specialist scientific literature makes it clear that methods of quantitative analysis and risk assessment form the basis for safety management in regard to WSS [8,9]. Results of hybrid risk-based decision making gain implementation by reference to pipe design parameters, with authors in work [10] prioritizing pipes in line with their rehabilitation needs. Comparisons of the use of a vulnerability assessment model based on Bayesian Belief Networks and the related uncertainty assessment model were in turn performed in [11], with respect to the emergency management of systems supplying drinking water.

Research on risk in technical systems includes the class of cognitive and practical methods of analysis and assessment, which constitute important elements of comprehensive research into their safety [12]. The following trends of safety research can be distinguished, as risk analysis, where assessment models form the basis for building decision models; and risk engineering, where the assessment of design options in terms of safety is the basis for choosing the best solution [13–15]. A water supply system's (WSS) loss of safety may result directly from the failure of its individual subsystems or elements, such as water intakes, pumping stations, the water distribution network (WDN) or its utilities [16–20]; from the failure of other technical systems (e.g., sewerage, energy, water structures) [21]; from undesirable extreme natural phenomena like floods and droughts; or from the incidental pollution of water sources [22,23]. Risk analysis of the WSS should be preceded by analysis of the reliability of all subsystems in terms of interruptions to water supply [24,25], as well as failure to meet quality requirements for health posing threats to consumers of water [26].

In water supply companies, records of breaks should be gathered, as these come to represent the basic source of information for reliability and safety analyses. Water supply companies should thus have guidelines about the way they collect information, also in the form of expert opinions, so that these can be used in the decision-making model [27–30]. The registers concerned should not only relate to the number of undesirable events, but should also contain precise data on the locations and causes of failure, their duration, repair time, possible consequences and other information required if the basic indicators used in the analysis of the safety of the WSS are to be determined [31–34], with these inter alia including the numbers of failures over the examined period of operation of the WSS, average working time without failure, average repair time, etc. [35–37]. Failures may be due to mistakes in design, construction or operation [38]. On the other hand, the causes of failure can be classified on the basis of the phenomena giving rise to them [39]. The formation of failures in a WDN is in fact a complex problem and each time requires a detailed analysis of the root causes [40–42]. Failures of the WDN can arise from errors directly attributable to activities of the designer, the contractor or the operator of the water supply system, through incorrect assumptions in the design and shortcomings at the stage of performance, mismatched material or improper service [43,44]. Also, the failure of a WDN may be caused by factory defective materials, defective sealing or anti-corrosive coatings [45,46]. There are also environmental causes reflecting meteorological conditions, including sudden changes in temperature and landslides, as well as causes resulting from the functioning of the WDN in combination with improper monitoring [47,48].

The process of risk analysis in regard to the operational safety of the WDN takes place via steps comprising of: determination of the WDN type, limiting values for rates of failure of water supply and the nuisance of performed repairs. These are followed by determination of the types of protection related to the functioning of the WDN, as well as risk levels, where these are assigned to the categories of tolerated, controlled and unacceptable risks [3].

The literature review shows that analysis based on failure rate is suggested for a WDN, as a proper tool for decision support when it comes to managing failures in such a system. The main objective of the work detailed here has then been the proposing of methods by which to identify and assess the risks associated with the functioning of a WSS in Poland, in relation to different implemented approaches. The results of the research allowed for the identification of areas and procedures able to mitigate risk of failure effectively.

2. Legal Regulations Regarding Safety of the Water Supply for Consumers

It is important to underline that the main task of water supply companies is to provide consumers with water of adequate quantity and quality which meets the requirements of the Drinking Water Directive (Council Directive 98/83/EC of 3 November 1998 on the quality of water intended for human consumption), with its latest amendments including Commission Directive (EU) 2015/1787 of 6 October 2015. In turn, in 2004, the World Health Organization (WHO) published guidelines for the application of water safety plans in which proposed analysis took in the chain of water supply intended for human consumption, inter alia, the water distribution network.

Analyses and assessments of WSS failure have been conducted in many countries since the second half of the last century [49–57], with obtained conclusions used to improve the programming, design, implementation and operation of the WSS; to inspire improvement and amendment of technical regulations, design standards, guidelines and instructions for the performance and acceptance of facilities; and to raise levels of technical knowledge and professional qualifications among designers, contractors and operators [58]. A very important issue is determination of the criterion value of risk, as a risk may be considered controlled when expected losses resulting from failure are acceptable to both a water company and the consumers of its water [2].

Risk acceptance criteria can be used in the decision-making process regarding the operation of such a system [3]. These criteria should take into account requirements relating to the reliability of subsystem operation, in terms of both quantity and quality, in accordance with applicable legal regulations, as well as social and economic conditions [59–61].

The management of the WSS should be improved constantly, through the pursuit of risk analysis and assessment and risk management in accordance with the recommendations of the WHO and existing regulations. The offer of water supply companies in terms of reliability of operation and water supply system safety should likewise improve constantly, notwithstanding changing economic and legal conditions, and not least in the circumstances of increasing pressure imposed by environmentalists and tougher quality standards relating to water for consumption [62,63].

For the purposes of Poland's Act on Crisis Management, a water supply system falls within the category of critical infrastructure, representing a system of key importance to the functioning of society and the state. As improper functioning of the above systems may pose a threat to human health or life, a high level of reliability and safety must be assured. The Crisis Management Act regulates the principles for the development of crisis management plans aimed at preventing crisis situations from arising, but also reacting should emergencies arise and ensuring the removal of consequences. Steps in the critical infrastructure protection process include risk assessment and indications as to priority activity, with a hierarchy being developed in line with the results of risk assessment carried out. The subject matter under consideration is thus of key importance in ensuring the proper functioning of a WSS.

3. Materials and Methods

3.1. The Failure Risk Approaches of Failure Framework and Data Source

The factors used in analyzing the failure rate among WDN were:

- type of a water pipe (mains, distribution or service connections),
- months in which the network was damaged.

The analysis presented represents a starting point for analysis of failure in a WDN. With this aim in mind, the research started with failure-rate analysis in respect of the aforementioned factors, with analysis of failure rates in line with risk acceptance criteria. To compare different criteria in the assessment of failure rates, and to establish the compatibility of expert assessments, a rank analysis was applied. After this, a proposal for modified Multi-Criteria Decision Analysis as regards assessment of the risk of failure in a WDN was presented.

Data regarding failures of the WDN cover a period of 13 years' operation, from 2005 to 2017. The analysis was carried out with Statsoft software [64]. Operational data were provided from the functioning of a real water supply system controlled by a municipal water company. Data on technical documentation and information obtained from the managers and services were also used.

3.2. Description of the Study Area

The distinguished water supply system is located in Poland's Subcarpathian province. The supplied city is located on the right bank of the Vistula, in south-eastern Poland, where it covers a total area of 86 km^2 and is located at 50°34′ N and 21°40′ E (Figure 1). The city has an urbanization rate of 555 inhabitant/km^2 and is at 160 m above sea level. The area within the city boundary is dominated by arable land, which accounts for some 5500 ha or as much as 65% of the total. About 3000 ha (36% of the area) is genuinely urban in character. The most limited form of cover (accounting for just 140 ha) in turn involves industrial areas, which thus represent no more than about 2% of the city. Climatically, this city is in the zone of lowlands and foothills. This denotes hot summers, relatively small amounts of rainfall (circa 600 mm) and winters that are not especially severe.

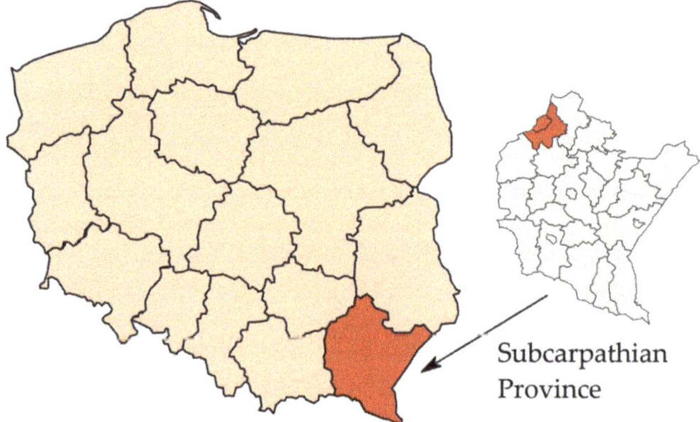

Figure 1. Location of the examined water supply system in Poland's Subcarpathian Province.

3.3. Characteristics of the Research Object

The urban water supply under study is currently used by the entire population of the distinguished city of some 51,000 people, as well as by some 3000 people living in other localities. Currently, the coverage by this water supply is close to 100%, with only single houses not being covered by the collective municipal water supply. In total, the length of the WDN is of 309.95 km.

The transmission of water from the clean water tanks takes place via two mains of diameters 400 and 500 mm. As the sections of mains connect twice in connection chambers, switch offs are possible in case of emergency. The water intake consists of two pumping intakes: the first consists of five drilling wells with a capacity of Q_{emax} = 183 m^3/h and the other consisting of 22 drilling wells with a capacity of Q_{emax} = 715 m^3/h. Expressed per day, this gives a value of 17,160 m^3/day. The volume of water taken from the intake is 2.92 million m^3/year. The Water Treatment Plant (WTP) is fed from a Quaternary aquifer with a free water table from a depth of about 15 m below ground level (BGL). The designed maximum capacity of the WTP is 715 m^3/h. The treated water flows from a clean tank to two water mains by means of three pumps. Two of these have a capacity of 440 m^3/h and a third one of 280 m^3/h. Their lifting capacity is 58 m.

The city's WDN consists of 16.8 km of mains and 191.7 km of distribution network. The water household connections network consists of 5000 connections and has a length of 100.85 km. The WDN is characterized by a mixed structure. The city center has a network with many loops, thanks to which, when a failure occurs, there is a possibility of residents continuing to be supplied with water thanks to a changing in the direction of water flow and inflow from another water line. The areas of suburban housing estates, in which there are single-family houses, are mostly supplied from a branched network. Investment activities in the area of construction and modernization of the WDN have been engaged in over the last few years, thanks to which more and more settlements are characterized by a looped system.

The largest share, about 43% of the entire WDN, is in the form of pipes less than 20 years old, but more than 11 years old. Approximately 90.2 km of the network is made of materials less than 10 years old, and this represents 31% of the entire WDN. The material structure of WDN is as follows: grey cast iron (37.7%), polyvinyl chloride (24.7%), galvanized steel (19.7%) and polyethylene (17.9%). The majority of house connections are made of cast iron and steel. It is only recently that pipes made of PVC (polyvinyl chloride) and PE (polyethylene) have been used. At present, no pipes forming the WDN are more than 40 years old. About 30.09% of the entire WDN is made up of pipes less than 10 years old, while about 42.7% is 11–20 years old and 26.4% of the entire WDN is above 20 years old.

3.4. Methods

3.4.1. Failure Rate Analysis

The analysis of failure rate in the WDN was performed by reference to the following formula [65,66]:

$$\lambda_j = \frac{n_j(t, t + \Delta t)}{N_j \times \Delta t}$$

(1)

where $n_j(t, t + \Delta t)$ is the number of all failures in the time interval Δt for the *j*th type of network; N_j is the length of the WDN of the *j*th type of network (km) and *j* is the type of WDN (e.g., for *M*—mains, *R*—distribution pipes, *SC*—service connections).

Standards as regards failure rate were proposed in the following works [6,67]. The former standards [6] provide that failure rates should not exceed criterion values as follows. In the case of mains λ_{Mcrit} is of less than 0.3 failures/(km^{-1}·year^{-1}), as compared with less than 0.5 for the distribution pipe λ_{Dcrit}, and ≤ 1.0 failures/(km^{-1}·year^{-1}) for service connections λ_{SCcrit}. In [67], proposals for the classification of criteria values of failure rates for the whole WDN were presented and classified in terms of reliability, with a high failure rate concerning low reliability when $\lambda_{lr\_crit} \geq$ 0.5 failures/(km^{-1}·year^{-1}), high reliability when $\lambda_{hr\_crit} \leq 0.1$ failures/(km^{-1}·year^{-1}), and average reliability between the criteria values mentioned $0.1 < \lambda_{ar\_crit} < 0.5$ failures/(km^{-1}·year^{-1}).

3.4.2. A Rank Analysis of Failure-Rate Criteria

To check the correlation between levels of compliance for individual criteria used to assess monthly failure rates, a rank analysis was applied, with its task being to determine correlations between two failure-rate criteria. Prior to the correlation calculation, data were sorted according to the degree of specific statistical characteristic, expressing them on ordinal scales, in line with the limit failure criteria given in Section 3.4.1.

The Spearman correlation coefficient is calculated as follows [1,68]:

$$r_S = 1 - \frac{6 \times \sum d_i^2}{n(n^2 - 1)}$$

(2)

where d_i is the difference between the two ranks of each observation and n the number of observations.

The failure rates for each month were ordered from the lowest to the highest for each year, with a rank assigned, whereby unity denotes the lowest failure rate and non-exceeded criterion of failure rate. The two separate assessments accorded with criteria presented in either [27] or [67].

3.4.3. Seasonal Analysis of Failure Rates

To assess variability due to the action of the factor of season, seasonal analysis was carried out after [69], by reference to:

- the seasonal index S_i given by:

$$S_i = \frac{\overline{y_i} \times d}{\sum\limits_{i=1}^{d} \overline{y_i}}, \tag{3}$$

where the arithmetic mean for failure rate in homonymous sub-periods (January, February,...., December) $(\overline{y_i})$ is:

$$\overline{y_i} = \frac{\sum\limits_{i=1}^{d} k_i}{12}, \tag{4}$$

where d is an annual cycle of fluctuations within which the monthly sub-periods were distinguished (d = 12, e.g., in January, February, etc.), and k_i the failure rate in a given month and year.

- absolute levels of seasonal fluctuations for individual sub-periods, calculated using:

$$g_i = S_i \times \overline{y} - \overline{y} \tag{5}$$

where g_i is the absolute level of seasonal fluctuation and $\overline{y}$ the average value given by:

$$\overline{y} = \frac{\sum \overline{y_i}}{d \times t}, \tag{6}$$

where t is the number of years of observation.

- the standard deviation (SD) characterizing absolute levels of seasonal variation, with these representing an assessment of variability due to the factor of season over a period of 13 years in operation:

$$SD(g_i) = \sqrt{\frac{\sum\limits_{i=1}^{d} g_i^2}{d}}. \tag{7}$$

With these dependent relationships taken account of, seasonal fluctuations in the occurrence of failures around WDNs were determined.

3.4.4. A Proposed Modified Multi-Criteria Decision Analysis Implementing an Analytic Hierarchy Process for Risk Assessment as Regards Failures in a WDN

Modified Multi-Criteria Decision Analysis (mMCDA) entails a choice of criteria influencing the risk of failure in a WDN, and the future occurrence thereof. The method suggested is based on risk-criteria grouping as regards failure in a WDN, with assessment then carried out by reference to determined points values under the Analytic Hierarchy Process (AHP) method [1,40,70,71]. It was assumed that risk means a measure by which to assess a hazard or threat resulting either from probable events beyond our control or from the possible consequences of a decision. Impacts are distinguished through the additive value of risk, which includes the category criterion of the size of possible financial losses resulting where failures arise. Evaluation criteria weights, as criterion of financial losses. The risk is interpreted by us in terms of expected losses.

In this way, the appropriate risk measure is calculated using Equation (8).

$$rA = \sum\limits_{j=1}^{m} w_j \times F_j(A), \tag{8}$$

where r is the additive value of risk, w_j the point weight for each subcategory criterion j relating to design, performance or operation, or social or financial environment or surroundings, where $j = \{1, 2, \ldots, m\}$, and F_j means category preference for alternatives taken into account by the method considered.

Depending on its influence on the risk index, each category was assigned a percentage weight that took account of importance in line with the Analytic Hierarchy Process (AHP), introduced by Thomas L. Saaty [72]. The procedure for using the AHP involves definition and analysis of the decision problem and goal-setting decisions. A set of criteria that have to be comparable are then established. The method also accepts criteria expressed quantitatively and qualitatively. Another important step in the procedure overall is the selection of relevant alternatives and the determination of implications of variations in the defined criteria. It is then possible to structure the hierarchical model. Input data are subject to comparisons of selected elements in pairs, with the advantage of one element over another determined in this way, in line with a nine-point pairwise comparison scale. The final decision is based on a synthesis of partial evaluation and selection of the best variant, through the creation of overall assessment scales using criteria and partial evaluation of alternatives. This relative preference is determined on a linear scale and verbally, taking into account the same significance or the strength of any advantage. Individual preferences and specific degrees of advantage in line with the Saaty scale are as shown in Table 1 [72,73].

Table 1. Scale of relative importance (after Saaty).

Interpretation		Value of a_{ij}	Definition
1	1	i and j are equally important	equal importance
3	1/3	i is slightly more important than j	moderate importance
5	1/5	i is more important than j	strong importance
7	1/7	i is far more important than j	very strong or demonstrated importance
9	1/9	i is absolutely more important than j	extreme importance
2, 4, 5, 8	1/2, 1/4, 1/5, 1/8	intermediate values	for comprise between the above values

Giving a relative preference means that the selection and evaluation of individual parameters are closer to reality. An expert's opinion is used to increase the substantive correctness of results. In the first place, the pairwise comparison of criteria allows them to be ordered qualitatively and as a matrix is being constructed in a quantitative way from the value of 1/9 to that of 9, with seventeen possible evaluations thus provided for.

In pairwise comparison by reference to an n by n matrix (A), so-called reversible pairwise comparisons are made, as already mentioned, and can be written as [72,73]:

$$A = [a_{ij}] \text{ for } i, j = 1, \ldots, n \tag{9}$$

where

$$a_{ij} = \frac{1}{a_{ji}}, \tag{10}$$

and

$$a_{ii} = 1 \tag{11}$$

the matrix is consistent, if:

$$a_{ij} = \frac{w_i}{w_j} \tag{12}$$

The matrix takes the following form:

$$A = \begin{bmatrix} 1 & a_{12} & \cdots & a_{1n} \\ \frac{1}{a_{12}} & 1 & \cdots & a_{2n} \\ \cdot & \cdot & \cdot & \cdot \\ \cdot & \cdot & \cdot & \cdot \\ \cdot & \cdot & \cdot & \cdot \\ \frac{1}{a_{1n}} & \frac{1}{a_{2n}} & \cdots & 1 \end{bmatrix} \tag{13}$$

As a result of calculations of the comparison matrix in pairs, vectors of the priorities $w = (w_1, \ldots, w_n)$, concerning the significance of elements, are obtained.

The obtained priority vectors can be presented in the form of a matrix of normalized grades, using Saaty's fundamental comparison scale:

$$
Aw = \begin{bmatrix} \frac{w_1}{w_1} & \frac{w_1}{w_2} & \cdots & \frac{w_1}{w_n} \\ \frac{w_2}{w_1} & \frac{w_2}{w_2} & \cdots & \frac{w_2}{w_n} \\ \cdot & \cdot & \cdot & \cdot \\ \cdot & \cdot & \cdot & \cdot \\ \cdot & \cdot & \cdot & \cdot \\ \frac{w_n}{w_1} & \frac{w_n}{w_2} & \cdots & \frac{w_n}{w_n} \end{bmatrix} \times \begin{bmatrix} w_1 \\ w_2 \\ \cdot \\ \cdot \\ \cdot \\ w_n \end{bmatrix} = n \times \begin{bmatrix} w_1 \\ w_2 \\ \cdot \\ \cdot \\ \cdot \\ w_n \end{bmatrix} \tag{14}
$$

hence equality occurs:

$$
Aw = nw \tag{15}
$$

To check if the matrix is consistent, the principal eigenvalue λ_{max} corresponding to the greatest value of the eigenvector, should be calculated in line with Equation (16):

$$
\lambda_{\max} = \frac{1}{w_i} \sum_{j}^{n} a_{ij} w_j, \tag{16}
$$

The matrix of pairwise comparisons $A = (a_{ij})$ is said to be consistent when its principal eigenvalue is close to n. To assess the compatibility of deviations, a Consistency Index (*CI*) is determined, in line with the dependence relationship:

$$
CI = \frac{\lambda_{\max} - n}{n - 1}, \tag{17}
$$

Next, a Consistency Ratio (*CR*) is determined, defining the degree to which comparison of the validity of characteristics is characterized by incompatibility. This is given by:

$$
CR = \frac{CI}{RI}, \tag{18}
$$

where *RI* is the Random Index, depending on the number of comparable items, according to the matrix dimensions listed in Table 2.

Table 2. *RI* values corresponding to matrix dimensions.

n	3	4	5	6	7	8	9	10	11	12
RI	0.52	0.89	1.11	1.25	1.35	1.40	1.45	1.49	1.52	1.54

To calculate weighting factors and consistency indexes, with a view to calculations performed being checked for their accuracy, use was made of methods involving the arithmetic and geometric means, as well as matrix multiplication. Depending on the designated consistency ratio, preferential information can be adopted or rejected. If the value of *CR* is greater than the permitted 0.1, an inconsistent matrix is present, and preferential information will need to be verified by decision makers. In such a case, actions will need to be taken to reduce low-quality observations and inconsistencies relating to them. It is very important that a decision problem taken up should not be too difficult and should be in line with assumptions adopted.

What is described here can combine with the experience and knowledge of experts to form a basis for expert opinions as regards the risk of failure. Indeed, the preparation of the categories and subcategories of criteria presented in Table 3 involved cooperation with designers, contractors and operators of the studied WDN. The key variables were also based on the relevant literature cited in the text. Expert opinion provides an opportunity for the experience of experts in a given field to be brought together, such that all the most important factors affecting the risk values associated with failures in a WDN may then be taken into account. Values for assessment subcategories of criteria are adopted in line with the importance of the damaged pipe, using a scale whereby a low hazard is on 1, a moderate hazard on 2 and a severe hazard on 3. If a given factor is not present in an analysis, a value of 1 for the subcategory is assumed.

Table 3 sets out proposed categories and subcategories of criteria for the analysis and identification of areas at risk of WDN failure [6,10,23,71].

For the mMCDA method a three-step scale to obtain the additive value of a risk category was adopted, risk assessment thus being provided for via implementation of AHP in line with a scale comprising tolerable risk (2.567 ÷ 3.06), controlled risk (3.06 ÷ 6.02) and unacceptable risk (6.02 ÷ 7.701). The proposed scale was based on the opinions of experts, who on the basis of their knowledge, experience and data in litteris assess risk categories. Assessment of the additive value of risk obtained supports the taking of decisions that relate to the operation or modernization of the system. Should a value indicating tolerable risk be obtained, no extra actions are required and the system is running in a proper and reliable way, with no preventive actions in the system needed either. Controlled risk means that the system is permitted to function but under the condition that modernization or renovation will commence. If an unacceptable level is found to have arisen, an immediate undertaking should be given, to the effect that the additive value of the risk category will be reduced. It should be noted that the consequences of failure to the water distribution subsystem are what cause periodic breaks in the delivery of water supply that is anyway of inadequate quality.

Table 3. Evaluation criteria weights.

No.			Categories and Subcategories of Criteria		Point Weighting of Subcategories
1	2	3	4		5
I	Ia)	Design	a reputable design office with a quality certificate, having completed projects on the reference list, and engaging in design using proven computer programs,		1
	Ib)		design office has certified designs and legitimizes itself by virtue of a list of references,		2
	Ic)		a person engaging in business activity, having in his portfolio partial projects for the expansion of WDN,		3
	Id)		network monitoring *	above-standard,	1
	Ie)			standard,	2
	If)			none,	3
	Ig)		corrosion protection	full,	1
	Ih)			standard,	2
	Ii)			none,	3
II	IIa)	Performance	construction company is certified and has completed investments in the reference list, procedures related to the acceptance of investments, pipes laid in accordance with the best available technology,		1
	IIb)		construction company has a reference list of completed investments, verification of material specifications and acceptance procedures is performed,		2
	IIc)		construction company is entering the water supply market, but has no experience in this area,		3
III	IIIa)	Operation	type of WDN according to higher priority to repair	SC,	1
	IIIb)			distribution network,	2
	IIIc)			mains,	3
	IIId)		failure rate, λ	to $0.5\ \mathrm{km}^{-1}\cdot\mathrm{year}^{-1}$,	1
	IIIe)			from $0.5\ \mathrm{km}^{-1}\cdot\mathrm{year}^{-1}$ to $1.0\ \mathrm{km}^{-1}\cdot\mathrm{year}^{-1}$,	2
	IIIf)			$>1.0\ \mathrm{km}^{-1}\cdot\mathrm{year}^{-1}$,	3
	IIIg)		dynamic loads, including difficulty of repairs in area in which a network is situated	pipeline in non-urbanized areas,	1
	IIIh)			pipeline in pedestrian traffic (under pavements),	2
	IIIi)			pipeline in the street,	3
	IIIj)		WDN age	to 20 years,	1
	IIIk)			from 20 to 50 years,	2
	IIIl)			above 50 years,	3
	IIIm)		WDN material	plastics,	1
	IIIn)			steel,	2
	IIIo)			grey cast iron,	3
IV	IVa)	Social	nuisance resulting from road occupation and green area	pipeline in non-urbanized area,	1
	IVb)			pipeline in pedestrian traffic,	2
	IVc)			pipeline in street,	3

Table 3. *Cont.*

No.			Categories and Subcategories of Criteria		Point Weighting of Subcategories
V	Va)	Financial	size of possible losses arising should failure occur	financial loss of up to 10^4 EUR,	1
	Vb)			financial loss from 10^4 to 10^5 EUR,	2
	Vc)			financial loss above 10^5 EUR,	3
	Vd)		difficulty of repairing damage	repair brigades are organized and equipped appropriately and are in full readiness for 24 h,	1
	Ve)			basic equipment to repair a failure, one-shift work,	2
	Vf)			lack of mechanized equipment to repair a failure,	3
VI	VIa)	Environment and surroundings	hydrogeological conditions	good,	1
	VIb)			average,	2
	VIc)			poor,	3
	VId)		density of underground infrastructure in the vicinity of the network	low,	1
	VIe)			average,	2
	VIf)			high.	3

Notes: * Above-standard network monitoring: full monitoring of the WDN through measurement of water pressure and flow rate, possession of specialized apparatus to detect water leaks by acoustic methods, unrestricted communication with the public via a phone line active 24 h a day, monitoring of water quality in a WDN by means of a protection and warning system. Standard: simplified monitoring of WDN by way of pressure measurement, inability to respond to minor leaks, water quality tests in WDN conducted. None: lack of monitoring of network and water quality.

4. Results and Discussion

4.1. Failure Rate Analysis

The failure risk analysis is found to contribute significantly to a reduction of failure overall, and a limiting of their consequences. The analysis question was based on operational data obtained over 13 years of observation. The λ_{WN} values for failure rates over the network as a whole (Figure 2), and for mains λ_M (Figure 3), distribution pipes λ_D (Figure 4), and SCs λ_{SC} (Figure 5), were determined in line with Equation (1). The failure rate for WDN pipes in subsequent months of the analyzed 2005–2017 period was then determined. To show seasonal fluctuations, Figure 6 was presented, with this supplying basic statistical characteristics for monthly values where failure rates are concerned. Results from seasonal calculations are as shown in Figure 7.

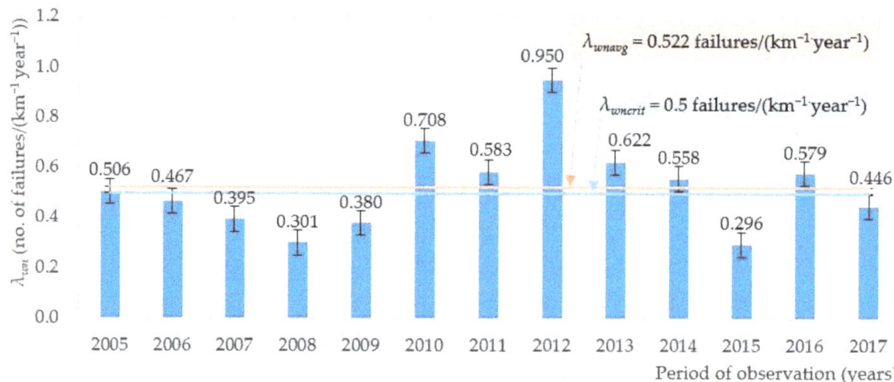

Figure 2. The failure rate for the whole network over the distinguished 2005–2017 time period.

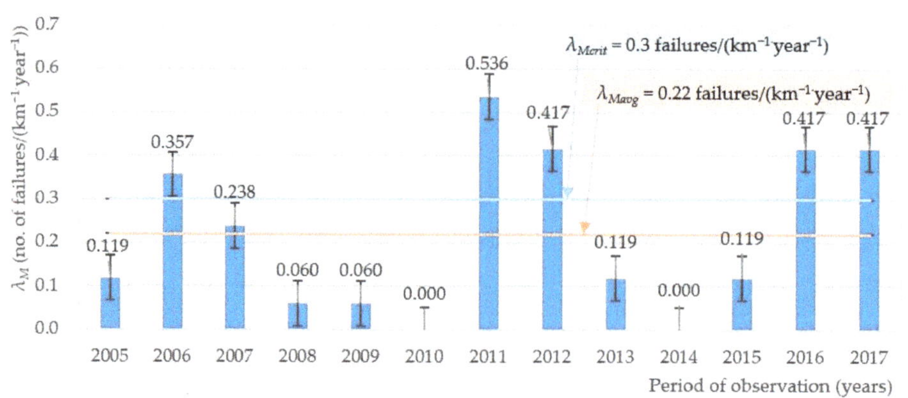

Figure 3. The failure rate for mains over the distinguished 2005–2017 period.

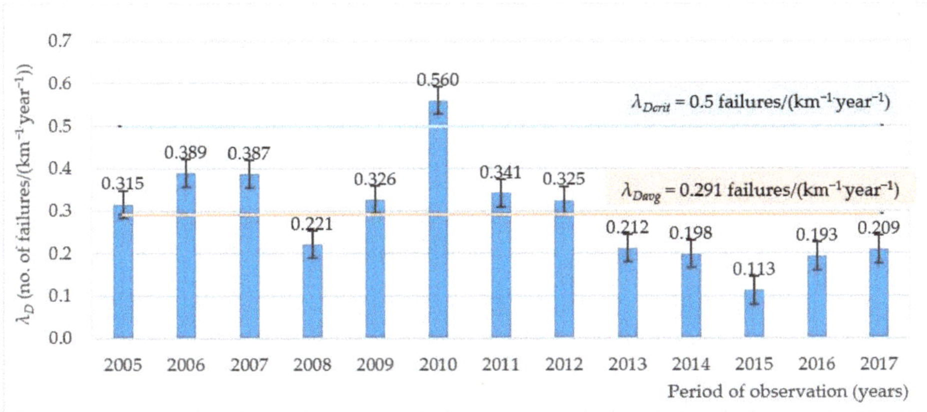

Figure 4. The failure rate for distribution pipes over the distinguished 2005–2017 period.

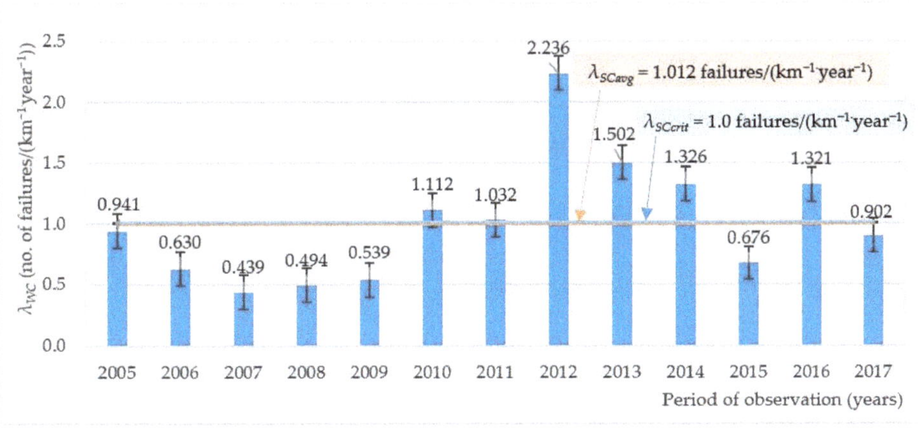

Figure 5. The failure rate for SCs over the distinguished 2005–2017 period.

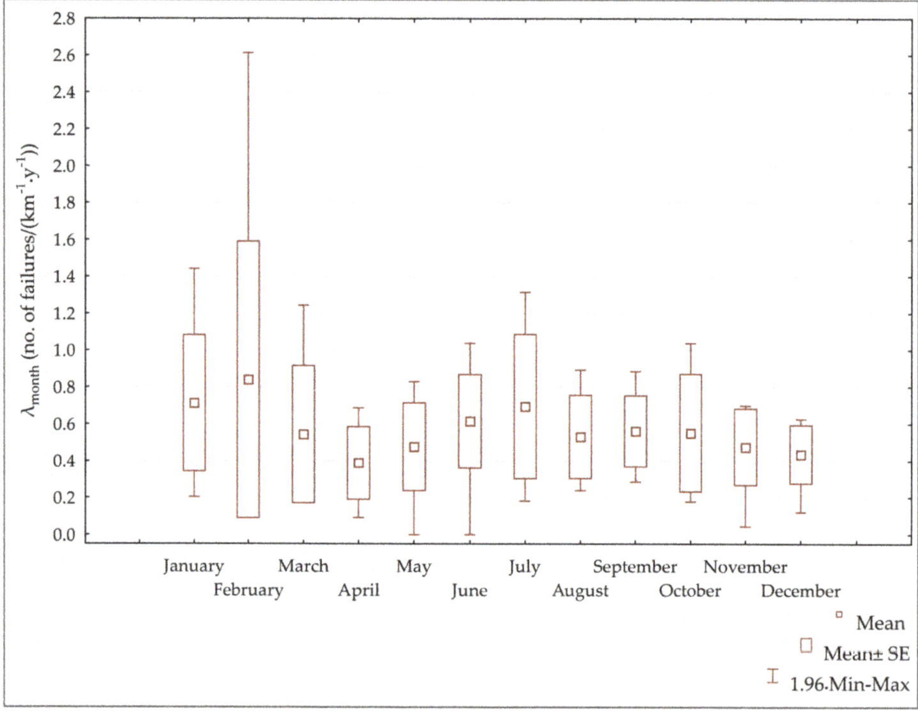

Figure 6. A box-whisker plot of the mean values for failure rates along the WDN in each month of the year and in line with the following operational data for the 2005–2017 period.

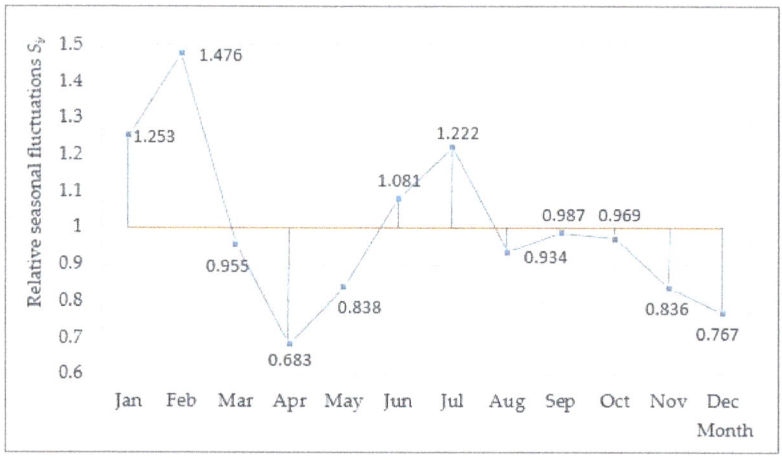

Figure 7. Relative seasonal fluctuations of failure rates in WDN.

Over the analyzed period of time, the average number of failures amounted to 1862 (on average 143 failures per year of observation), while the corresponding average failure rate is 0.522 failures/(km^{-1}·year^{-1}). Before the analysis was commenced, verification of operational data was performed, in respect of the classification of data on the failure rate of individual types of pipes and changes resulting from repairs and modernizations. The determined SD for the whole network was of 0.171 failures/(km^{-1}·year^{-1}), as compared with 0.178 failures/(km^{-1}·year^{-1}) for the mains, 0.113 for distribution pipes, and 0.483 failures/(km^{-1}·year^{-1}) for SCs. Where the whole network is taken into account (in line with the data presented in Figure 2), the highest failure rates were found to have occurred in 2010 and 2012, with 199 and 260 failures. The lowest failure rates occurred in 2008 and 2015, with 78 and 85 failures respectively. The highest failure rate in 2010 may be related to flooding, which took place twice that year and gave rise to consequences as regards liquidation that caused the number of failures to increase. The detailed analysis in turn indicated that the high failure rate in 2012 reflected low temperature, which caused the freezing of SC. Dependence is observed in failure-rate distribution over the month-long span of time in the analyzed years, and this is primarily seen to be due to changes in ground temperature in the winter-spring period. In 2012, about 24% of all failures (about 63 in absolute terms) occurred in February, with this mainly reflecting changes in ground temperature during periods of the winter season. An increased number of failures also occurs in July, with this probably owing to heat-driven drying of soil leaving pipes more prone to cracking. Such a situation arises with the superficial laying of pipes and low water flow at night then causing the freezing of water in the pipes more prone to frost. Over the last few years, the length of distribution pipes and SCs has increased significantly. Since 2005, about 22.2 km of water-supply connections and 45.8 km of distribution pipes have been constructed. It is worth underlining that failures of the main network do not cause a break in water supply because the mains consists of two pipelines. According to data presented (Figures 3 and 5), water supply connections (λ_{SCavg} = 1.012 failures/(km^{-1}·year^{-1}) are most often damaged, while the main network is least often damaged (λ_{Mavg} = 0.22 failures/(km^{-1}·year^{-1}). The occurrence of more failures in distribution pipes and SCs is related to the age and type of material from which they are made. Over the years, the failure rate of distribution pipes has declined; attesting to modernization work on the WDN in the city. According to the data presented in Figure 4, in recent years there has been a tendency for the failure rate along distribution pipes to decrease. The significant reduction in failure rate occurred after 2010, it was due to the start of the modernization of the WDN. The distribution network meets the criteria that the failure rate should be less than 0.5 failures/(km^{-1}·year^{-1}), except in the year 2010 when the criterion value of about 12% was slightly exceeded. In the case of SC there was a decrease in the failure rate after 2012. In that period SCs have met the standards ($\lambda_{SC} \leq$ 1.0 failure/(km^{-1}·year^{-1}), and, as was mentioned above, this reflects modernization and a change in the material from which the network is made, given the use of the PVC and PE pipes less prone to failure and characterized by lower failure rates than pipes made from steel or cast iron. Taking into account the criteria presented in work [67] and the mean failure rate, the network as a whole is characterized by low reliability, with $\lambda \geq$ 0.5 failures/(km^{-1}·year^{-1}), while distribution pipes and SCs are of average reliability, with mean values for failure rates between 0.1 and 0.5 failures/(km^{-1}·year^{-1}). The results of calculations show that, in every year of the analyzed period, due to seasonal fluctuations (as Figure 6 shows), the total number of failures in February was higher than the monthly average (100%) by approximately 47.6%, in January by 25.3% and in July by 22.2%. In turn, in April the value is lower than the average by 31.7%, and so on (Figure 7). Such tendencies are also present in other systems [44]. Absolute seasonal fluctuation in g_i informs us that, in January, the total number of failures is higher than in an average month (at $\bar{y}$ = 0.569 failures/(km^{-1}·year^{-1})), and therefore by 0.271 failures/(km^{-1}·year^{-1}). In turn, in April, the figure is lower than the average by 0.18 failures/(km^{-1}·year^{-1}) (Figure 8). SD for absolute seasonal variation levels, over the period of 13 years in operation is at the level of 0.123 failures/(km^{-1}·year^{-1}).

The box-whisker plot for mean values of failure rates allowed for assessment overall, and the most marked variation was shown for the winter month of January, which is characterized by a SD equal to 0.75. The most limited differentiation in turn occurs in the summer month of July, for which the SD value is at the level of 0.39. The detailed analysis in the last year of observation indicates that the highest indicator values for failure rate characterizes pipes made of steel, representing 0.23 failures/(km^{-1}·year^{-1}), as compared with the lowest failure rate determined for PE, at the level of about 0.09 failures/(km^{-1}·year^{-1}).

The Spearman's rank correlation coefficient was used to compare the assessment of the different criteria in regard to each monthly failure rates. When the obtained values for coefficients were compared, a high level of agreement regarding the assessment and classification of failure rate was only found in the case of mains (at the level of 0.589, and hence significant at $p < 0.05$). In the case of service connections, the average value achieved for the relationship was of 0.395. The lack of failure rates below 0.1 failures/(km^{-1}·year^{-1}), caused the same assessment according to two criteria for the whole network and for the distribution pipes, and there was also functional dependence noted between these two assessments.

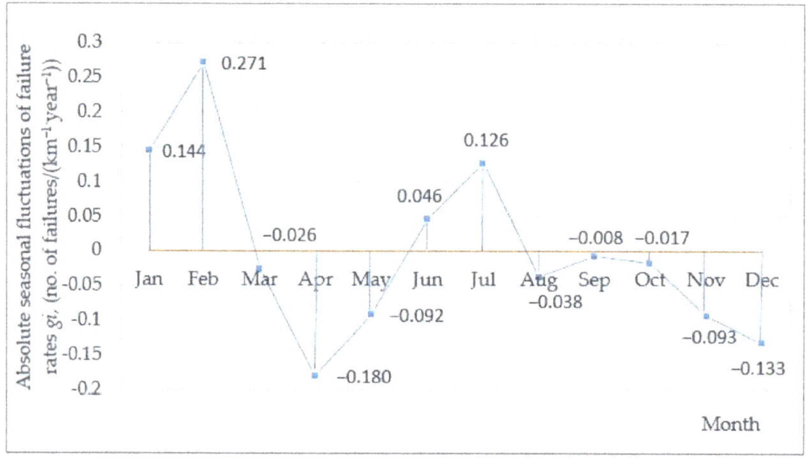

Figure 8. Absolute seasonal fluctuations of failure rates in WDN, in (no. of failures/(km^{-1}·year^{-1})).

4.2. Results of Failure Risk Assessment for WDN

The performed analysis focused in on a section of the distribution network along with traffic difficulties caused by repairs of the WDN which arise very often. The examined section is of total length 2.4 km, and forms part of the WDN in the city under consideration supplying 50,000 inhabitants. The expert analysis revolved around six main criteria, which were compared in terms of pairs of individual criteria, in line with the scale developed that is presented in Table 4.

Table 4. Matrix construction and weighting calculation for categories associated with the failure risk assessment.

Category	I	II	III	IV	V	VI	Weight
I	1	2	2	4	4	3	0.3262
II	0.5	1	2	4	4	3	0.2589
III	0.5	0.5	1	4	3	3	0.1959
IV	0.25	0.25	0.25	1	2	2	0.0856
V	0.25	0.25	0.333	0.5	1	2	0.0712
VI	0.333	0.333	0.333	0.5	0.5	1	0.0622
Total	2.833	4.333	5.917	14	14.5	14	0.3262
$\lambda_{max} = 6.3071; CI = 0.0614; RI = 1.25; CR = 0.0491$							

As the determined *CR* is acceptable (<0.1), the results obtained may be adopted. The preference vectors determined on this basis are shown in Table 5 [72].

Table 5. A characterisation of the network.

No.			Categories and Subcategories of Criteria		Weight	
					Point Weight of Subcategories	Categories
1	2	3	4		5	6
I	Ib)	Designing		design office has certified designs and legitimizes itself by virtue of a list of references,	2	0.3262
	Ie)		network monitoring	standard, simplified monitoring of water- WDN with the use of pressure measurement,	2	0.3262
	Ii)		corrosion protection	not applicable	3	0.3262
II	IIb)	Performance	construction company has a reference list of completed investments, verification of material specifications and acceptance procedures is performed,		2	0.2589
III	IIIb)	Operation	type of WDN	distribution network,	2	0.1959
	IIId)		failure rate, λ	to 0.5 $\mathrm{km}^{-1} \cdot \mathrm{year}^{-1}$,	1	0.1959
	IIIg)		dynamic loads	pipeline in not urbanized areas,	1	0.1959
	IIIk)		WDN age	from 20 to 50 years,	2	0.1959
	IIIm)		WDN material	plastics,	1	0.1959
IV	IVa)	Social	nuisance of road and green area users	pipeline in non-urbanized area,	1	0.0856
V	Va)	Financial	size of possible losses arising should failure occur	financial loss of up to 10^4 EUR,	1	0.0712
	Ve)		difficulty of repairing damage	basic equipment to repair a failure, one-shift work,	2	0.0712
VI	VIc)	Environment and surroundings	hydrogeological conditions	poor,	3	0.0622
	VId)		density of underground infrastructure in the vicinity of the network	low.	1	0.0622

According to Table 2, the value of $r(A)$ was 4.72, which corresponds to the category of a controlled risk level. The obtained value indicates that some improvement in the work of the WDN, or repair of certain sections of the WDN, should be considered to minimize risk of failure. Directions proposed for the WDN operator responsible for the monitoring of proper functioning Included the introduction of preventive measures, with a view to conditions leading directly to a WDN threat not arising. At this stage, the possibilities of undesirable events occurring should be analyzed from the point of view of the reliability of the water supply system, through water reserving, and the monitoring of failure around the WDN–as can be achieved by introducing repair brigades that are properly organized and fully ready 24 h a day. Introduction of a satisfactory level of protection, with a view to protecting the consumers of drinking water using the public WSS, also seeks to minimize possible losses. Minimizing the effects of unintentional events is also advocated, with water taken from alternative sources if necessary and the process of informing the public about interruptions in water supply checked for its effectiveness.

5. Conclusions and Perspectives

The issues presented may prove of great interest to authorities managing the operation of a WSS in water supply companies, as well as to the scientific community, doctoral students and students of technical and agricultural universities, employees of design companies and people who deal with water treatment technologies, and the operation of WDN and internal installations.

The methodology in question can gain direct application even in relation to water supply systems, given the relative universality of the approach presented here. Moreover, the presented method by which to pursue failure analysis should be a major element of any comprehensive management. If numerous failures arise in a WDN, the water company is obliged to modernize and renovate, so as to minimize pipeline failures. In its initial stages, managing of WDN safety entails creating a database of weak points of system functioning, with particular emphasis placed on frequency of occurrence and associated negative effects. In safety management, decisions are made as regards the selection of risk protection measures, implementation into operational practice and checks needing to be carried out in regard to the effectiveness of applied solutions.

Further work in this area will be focused on researching exposure to failure using operational data, field investigations and analyses of experts, with this facilitating identification of particular areas with high and unacceptable failure rates, in this way achieving a classification of sections of the network risk-mapped as in need of renovation. A further important aspect will entail analysis of failure assessment in the context of the determination of acceptance criteria vis-à-vis failure. The suggested method of failure risk assessment for WDNs is among expert methods that may be deployed, and may form part of the decision making process as regards plans for the modernization of a WSS. It can also be used where operational data prove insufficient and can be combined with the experience and knowledge of experts as a basis for opinions regarding the risk of failure. Furthermore, this type of analysis is very helpful in classifying those segments of a WDN that need repairing. The multi-criteria methods in failure risk assessment involve cooperation between designers, contractors and operators of the WDN, so this denotes an opportunity to combine the experience of experts in given fields. The methods proposed here from work based on operational data and expert knowledge can form a basis for a comprehensive risk management program that forms part of the water safety plans recommended by the WHO, which will be obligatory, given the need for modern standards regarding the safety of drinking water to be met.

Author Contributions: All authors equally contributed to the development of this manuscript.

Funding: This research was funded by Faculty of Civil and Environmental Engineering and Architecture, Rzeszow University of Technology, 35-959 Rzeszow, Poland.

Conflicts of Interest: The authors declare no conflict of interest.

References

1. Rak, J.R.; Tchórzewska-Cieślak, B. *Risk in Operation of Collective Water Supply Systems*; Publishing House of Seidel Przywecki: Warsaw, Poland, 2013.
2. Rak, J. *Safety of Water Supply System*; Polish Academy of Science: Warsaw, Poland, 2009.
3. Rak, J.; Tchórzewska-Cieślak, B. *Risk Factors in the Operation of Water Supply Systems*; Publishing House of Rzeszow University of Technology: Rzeszow, Poland, 2007.
4. Rak, J. *Fundamentals of Water Supply System Safety*; Polish Academy of Science: Lublin, Poland, 2005.
5. Rak, J. A study of the qualitative methods for risk assessment in water supply systems. *Environ. Prot. Eng.* **2003**, *29*, 123–133.
6. Rak, J.R. Some aspects of risk management in waterworks. *Ochr. Sr.* **2007**, *29*, 61–64.
7. Rak, J.; Tchórzewska-Cieślak, B. The possible use of the FMEA method to ensure health safety of municipal water. *J. Konbin* **2010**, *3*, 143–154. [CrossRef]
8. Rak, J.R. Methods of reliability index determination concerning municipal water quality. *J. Konbin* **2008**, *5*, 157–173. [CrossRef]
9. Tchórzewska-Cieślak, B. *Methods for Analysis and Assessment of Risk of Failure in Water Distribution Subsystem*; Publishing House of Rzeszow University of Technology: Rzeszow, Poland, 2011.
10. Salehi, S.; Tabesh, M.; Ghazizadeh, M.J. HRDM method for rehabilitation of pipes in water distribution networks with inaccurate operational-failure data. *J. Water Res. Plan. Manag.* **2018**, *144*. [CrossRef]
11. Pagano, A.; Pluchinotta, I.; Giordano, R.; Petrangeli, A.B.; Fratino, U.; Vurro, M. Dealing with uncertainty in decision-making for drinking water supply systems exposed to extreme events. *Water Res. Manag.* **2018**, *32*, 2131–2145. [CrossRef]
12. Tchorzewska-Cieslak, B.; Rak, J. Method of identification of operational states of water supply system. Environmental Engineering III. In Proceedings of the 3rd Congress of Environmental Engineering, Pafos, Cyprus, 13–16 September 2009; pp. 521–526.
13. Izquierdo, J.; López, P.A.; Martinéz, F.A.; Pérez, R. Fault detection in water supply systems using hybrid (theory and data-driven) modeling. *Math. Comput. Model.* **2007**, *46*, 341–350. [CrossRef]
14. Khomsi, B.D.; Walters, G.A.; Thorley, A.R.D.; Ouazar, D. Reliability tester for water distribution networks. *J. Comput. Civ. Eng.* **1996**, *10*, 10–19. [CrossRef]

15. Tchorzewska-Cieslak, B.; Wlodarczyk-Makula, M.; Rak, J. Safety analysis of the wastewater treatment process in the field of organic pollutants including PAHs. *Desalin. Water Treat.* **2017**, *72*, 146–155. [CrossRef]

16. Bergel, T.; Pawełek, J. Quantitative and economical aspects of water loss in water-pipe networks in rural areas. *Environ. Prot. Eng.* **2008**, *34*, 59–64.

17. Zimoch, I. Pressure control as part of risk management for a water-pipe network in service. *Ochr. Sr.* **2012**, *34*, 57–62.

18. Kuliczkowski, A.; Kuliczkowska, E.; Kubicka, U. Design of the pipelines considering exploitative parameters. Underground Infrastructure of Urban Areas. In Proceedings of the International Conference on Underground Infrastructure of Urban Areas, Wroclaw, Poland, 22–24 October 2008; pp. 165–170.

19. Pozos-Estrada, O.; Sanchez-Huerta, A.; Brena-Naranjo, J.A.; Pedrozo-Acuna, A. Failure analysis of a water supply pumping pipeline system. *Water* **2016**, *8*. [CrossRef]

20. Wąsik, E.; Chmielowski, K.; Cupak, A.; Kaczor, G. Stability monitoring of the nitrification process: Multivariate statistical analysis. *Pol. J. Environ. Stud.* **2018**, *27*, 2303–2313. [CrossRef]

21. Królikowska, J. Application of PHA method for assessing risk of failure on the example of sewage system in the City of Krakow. *Rocz. Ochr. Sr.* **2011**, *13*, 693–710.

22. Berardi, L.; Ugarelli, R.; Rostum, J.; Giustolisi, O. Assessing mechanical vulnerability in water distribution networks under multiple failures. *Water Resour. Res.* **2014**, *50*, 2586–2599. [CrossRef]

23. Kozłowski, E.; Mazurkiewicz, D.; Kowalska, B.; Kowalski, D. Application of a multidimensional scaling method to identify the factors influencing on reliability of deep wells. *Adv. Intell. Syst.* **2019**, *835*, 56–65. [CrossRef]

24. Tchórzewska-Cieślak, B.; Boryczko, K.; Eid, M. Failure scenarios in water supply system by means of fault tree analysis. In *Advances in Safety, Reliability and Risk Management*; Taylor & Francis Group: Abingdon, UK, 2012; pp. 2492–2499.

25. Zimoch, I. Method of risk analysis and assessment of water supply system exploitation in agricultural areas. *Ochr. Sr.* **2016**, *38*, 33–38.

26. Puri, D.; Borel, K.; Vance, C.; Karthikeyan, R. Optimization of a water quality monitoring network using a spatially referenced water quality model and a genetic algorithm. *Water* **2017**, *9*. [CrossRef]

27. Jung, D.; Yoo, D.; Kang, D.; Kim, J. Linear model for estimating water distribution system. *J. Water Res. Plan. Manag.* **2016**, *142*, 04016022. [CrossRef]

28. Moglia, M.; Davis, P.; Burn, S. Strong exploration of a cast iron pipe failure model. *Reliab. Eng. Syst. Saf.* **2008**, *93*, 863–874. [CrossRef]

29. Wang, Y.; Zayed, T.; Moselhi, O. Prediction models for annual break rates of water mains. *J. Perform. Constr. Fac.* **2009**, *23*, 47–54. [CrossRef]

30. Zimoch, I.; Łobos, E. Application of the theil statistics to the calibration of a dynamic water supply model. *Environ. Prot. Eng.* **2010**, *36*, 105–115.

31. Mamo, T.G. Risk-based approach to manage aging urban water main infrastructure. *J. Water Supply Res. Technol.* **2015**, *64*, 260–269. [CrossRef]

32. Shahata, K.; Zayed, T. Data acquisition and analysis for water main rehabilitation techniques. *Struct. Infrastruct. E* **2012**, *8*, 1054–1066. [CrossRef]

33. Scheidegger, A.; Scholten, L.; Maurer, M.; Reichert, P. Extension of pipe failure models to consider the absence of data from replaced pipes. *Water Res.* **2013**, *11*, 3696–3705. [CrossRef] [PubMed]

34. Tabesh, M.; Soltani, J.; Farmani, R.; Savic, D. Assessing pipe failure rate and mechanical reliability of water distribution networks using data-driven modeling. *J. Hydroinform.* **2009**, *11*, 1–17. [CrossRef]

35. D'Ercole, M.; Righetti, M.; Raspati, G.S.; Bertola, P.; Ugarelli, R.M. Rehabilitation planning of water distribution network through a reliability-based risk assessment. *Water* **2018**, *10*. [CrossRef]

36. Iwanejko, R.; Bajer, J. Determination of the optimum number of repair units for water distribution systems. *Arch. Civ. Eng.* **2009**, *55*, 87–101.

37. Martinez-Codina, A.; Castillo, M.; Gonzales-Zeas, D.; Garrote, L. Pressure as a predictor of occurrence of pipe breaks in water distribution networks. *Urban Water J.* **2016**, *13*, 676–686. [CrossRef]

38. Królikowska, J. Damage evaluation of a town's sewage system in southern poland by the preliminary hazard analysis method. *Environ. Prot. Eng.* **2011**, *37*, 131–142.

39. Szpak, D.; Tchórzewska-Cieslak, B. Assessment of the failure rate of water supply system in terms of safety of critical infrastructure. *Chemik* **2014**, *6*, 862–867.

40. Darvini, G. Comparative analysis of different probability distributions of random parameters in the assessment of water distribution system reliability. *J. Hydroinform.* **2014**, *14*, 272–287. [CrossRef]

41. Boxall, J.B.; O'Hagan, A.; Pooladsaz, S.; Saul, A.J.; Unwin, D.M. Estimation of burst rates in water distribution mains. *Proc. Inst. Civ. Eng.-Water Manag.* **2007**, *160*, 73–82. [CrossRef]

42. Shin, H.; Kobayashi, K.; Koo, J.; Do, M. Estimating burst probability of water pipelines with a competing hazard model. *J. Hydroinform.* **2016**, *18*, 126–135. [CrossRef]

43. Arai, Y.; Koizumi, A.; Inakazu, T.; Watanabe, H.; Fujiwara, M. Study on failure rate analysis for water distribution pipelines. *J. Water Supply Res. Technol.* **2010**, *59*, 429–435. [CrossRef]

44. Hotloś, H. *Quantitative Assessment of the Effect of Some Factors on the Parameters and Operating Costs of Water-Pipe Networks*; Wrocław University of Technology Publishing House: Wrocław, Poland, 2007.

45. Estokova, A.; Ondrejka Harbulakova, V.; Luptakova, A.; Stevulova, N. Performance of fiber-cement boards in biogenic sulphate environment. *Adv. Mater. Res.* **2014**, *897*, 41–44. [CrossRef]

46. Hotlos, H. Quantitative assessment of the influence of water pressure on the reliability of water-pipe networks in service. *Environ. Prot. Eng.* **2010**, *36*, 103–112.

47. Bogardi, I.; Fülöp, R. A Spatial probabilistic model of pipeline failures. *Period Polytech.-Civ.* **2011**, *55*, 161–168. [CrossRef]

48. Debón, A.; Carrión, A.; Cabrera, E.; Solano, H. Comparing risk of failure models in water supply networks using ROC curves. *Reliab. Eng. Syst. Saf.* **2010**, *95*, 43–48. [CrossRef]

49. Kleiner, Y.; Rajani, B. Comprehensive review of structural deterioration of water mains: Statistical models. *Urban Water J.* **2001**, *3*, 131–150. [CrossRef]

50. Kowalski, D.; Kowalska, B.; Kwietniewski, M. Monitoring of water distribution system effectiveness using fractal geometry. *Bull. Pol. Acad. Sci.* **2015**, *63*, 155–161. [CrossRef]

51. Kutyłowska, M. Comparison of two types of artificial neural networks for predicting failure frequency of water conduits. *Period Polytech.-Civ.* **2017**, *61*, 1–6. [CrossRef]

52. Kwietniewski, M.; Rak, J. *Reliability of Water and Wastewater Infrastructure in Poland*; Polish Academy of Science: Warsaw, Poland, 2010.

53. Mays, W.L. *Reliability Analysis of Water Distribution Systems*; American Society of Civil Engineers: New York, NY, USA, 1998.

54. Pelletier, G.; Mailhot, A.; Villeneuve, J.P. Modeling water pipe breaks-three case studies. *J. Water Res. Plan. Manag.* **2003**, *129*, 115–123. [CrossRef]

55. Shamir, U.; Howard, C.D.D. An analytical approach to scheduling pipe replacement. *J. Am. Water Works Assoc.* **1979**, *71*, 248–258. [CrossRef]

56. Toumbou, B.; Villeneuve, J.P.; Beardsell, G.; Duchesne, S. General model for water-distribution pipe breaks: Development, methodology, and application to a small city in Quebec, Canada. *J. Pipeline Syst. Eng.* **2014**, *5*, 04013006. [CrossRef]

57. Xu, Q.; Chen, Q.; Li, W. Application of genetic programming to modeling pipe failures in water distribution systems. *J. Hydroinform.* **2011**, *13*, 419–428. [CrossRef]

58. Shuang, Q.; Liu, Y.S.; Tang, Y.Z.; Liu, J.; Shuang, K. System reliability evaluation in water distribution networks with the impact of valves experiencing cascading failures. *Water* **2017**, *9*. [CrossRef]

59. Rak, J.R.; Kucharski, B. Sludge management in water treatment plants. *Environ. Prot. Eng.* **2009**, *35*, 15–21.

60. Walski, T.M.; Pelliccia, A. Economic analysis of water main breaks. *J. Am. Water Works Assoc.* **1982**, *74*, 140–147. [CrossRef]

61. Tchórzewska-Cieślak, B.; Pietrucha-Urbanik, K.; Urbanik, M.; Rak, J.R. Approaches for safety analysis of gas-pipeline functionality in terms of failure occurrence: A case study. *Energies* **2018**, *11*, 1589. [CrossRef]

62. Nowacka, A.; Wlodarczyk-Makula, M.; Tchorzewska-Cieslak, B.; Rak, J. The ability to remove the priority PAHs from water during coagulation process including risk assessment. *Desalin. Water Treat.* **2016**, *57*, 1297–1309. [CrossRef]

63. Seo, J.; Koo, M.; Kim, K.; Koo, J. A study on the probability of failure model based on the safety factor for risk assessment in a water supply network. *Procedia Eng.* **2015**, *119*, 206–215. [CrossRef]

64. StatSoft, Inc. STATISTICA (Data Analysis Software System), Version 12. Available online: www.statsoft.com (accessed on 15 July 2018).

65. Kwietniewski, M.; Roman, M.; Kłos-Trębaczkiewicz, H. *Relaibility of Water Supply and Sewage Systems*; Arkady: Warsaw, Poland, 1993.

66. Wieczysty, A. *Reliability of Water and Wastewater Systems I and II*; Publishing House of Cracow University of Technology: Kraków, Poland, 1990.

67. Kwietniewski, M. Failure of water supply and wastewater infrastructure in Poland based on the field tests. In Proceedings of the XXV Scientific-Technical Conference, Międzyzdroje, Poland, 24–27 May 2011; pp. 12–140.

68. Corder, G.W.; Foreman, D.I. *Nonparametric Statistics: A Step-by-Step Approach*; Wiley: Hoboken, NJ, USA, 2014.

69. Sobczyk, M. *Statistics: Theoretical and Practical Aspects*; Maria Curie-Sklodowska University (UMCS): Lublin, Poland, 2006.

70. Figueira, J.; Greco, S.; Ehrgott, M. *Multicriteria Decision Analysis: State of the Art Surveys*; Springer: New York, NY, USA, 2005. [CrossRef]

71. Tchórzewska-Cieślak, B.; Pietrucha-Urbanik, K. Methods for integrated failure risk analysis of water network in terms of water supply system functioning management. *J. Pol. Saf. Reliab. Assoc.* **2017**, *8*, 157–166.

72. Saaty, L.T. A scaling method for priorities in hierarchical structures. *J. Math. Psychol.* **1977**, *15*, 234–281. [CrossRef]

73. Saaty, L.T. *The Analytic Hierarchy Process*; McGraw-Hill: New York, NY, USA, 1980.

© 2018 by the authors. Licensee MDPI, Basel, Switzerland. This article is an open access article distributed under the terms and conditions of the Creative Commons Attribution (CC BY) license (http://creativecommons.org/licenses/by/4.0/).

Article

Flow Velocity Distribution Towards Flowmeter Accuracy: CFD, UDV, and Field Tests

Mariana Simão [1,*], Mohsen Besharat [1], Armando Carravetta [2] and Helena M. Ramos [1]

[1] CERIS, Instituto Superior Técnico, Universidade de Lisboa, 1049-001 Lisboa, Portugal; mohsen.besharat@tecnico.ulisboa.pt (M.B.); helena.ramos@tecnico.ulisboa.pt (H.M.R.)

[2] Department of Hydraulic, Geotechnical and Environmental Engineering, Università di Napoli Federico II, via Claudio, 21, Napoli 80125, Italy; arcarrav@unina.it

* Correspondence: m.c.madeira.simao@tecnico.ulisboa.pt; Tel.: +351-218-418-151

Received: 22 October 2018; Accepted: 6 December 2018; Published: 8 December 2018

Abstract: Inconsistences regarding flow measurements in real hydraulic circuits have been detected. Intensive studies stated that these errors are mostly associated to flowmeters, and the low accuracy is connected to the perturbations induced by the system layout. In order to verify the source of this problem, and assess the hypotheses drawn by operator experts, a computational fluid dynamics (CFD) model, COMSOL Multiphysics 4.3.b, was used. To validate the results provided by the numerical model, intensive experimental campaigns were developed using ultrasonic Doppler velocimetry (UDV) as calibration, and a pumping station was simulated using as boundary conditions the values measured in situ. After calibrated and validated, a new layout/geometry was proposed in order to mitigate the observed perturbations.

Keywords: experiments; ultrasonic Doppler velocimetry (UDV); flowmeters; computational fluid dynamics (CFD); pipe system efficiency

1. Introduction

Worldwide, water companies use several flowmeters to measure the amount of water distributed. This equipment is quite vital for the management of water companies since important improvements are made according to the data provided by the available measures. The data provided by these devices also influences several performance indicators regarding the management of the system, such as non-revenue water (NRW) and water balances. The NRW is the volume of treated water that is not purchased. The water balances are important tools to detect leaks throughout the supply and distribution processes. Both these tools require accurate measurements [1].

The flow measurements can be correlated to the system efficiency. Usually, the systems are in part driven by gravity and by pressure differences, which require a pumping station. If the measurement accuracy is guaranteed, a higher energy efficiency level is possible to achieved, making possible a working period plan in the lower energy tariffs depending on the regularization ability and the water needs downstream [2,3].

Thus, the correct measurement by using flowmeters is an extremely important factor in terms of hydraulic system accuracy and management. Amongst other domains, the measured flow values are one of the key issues to detect leaks in pipe systems. Therefore, for any hydraulic system manager, the accuracy in flow measurements has significant impacts regarding both planning and investment decisions. Even when following all the constraints imposed by manufactures, incongruences can be identified in the field [4]. These irregularities suggested many times the occurrence of leaks in pipe systems. This conclusion raises a new problem that was not yet well identified [5].

Hence, this research has the objective to analyze different variables, namely the pipe layout and the way it can disturb the flow and flowmeter accuracy. The purpose is to assess how vertical or horizontal curves, expansions and reductions, among other geometries can influence the flow and, consequently the measurements. To fulfil this goal, the influence of several perturbations were identified, using an electromagnetic flowmeter and ultrasonic Doppler velocimetry (UDV), and compared with the computed velocity profiles, using CFD models. The numerical model was calibrated and validated using the same conditions as the experimental facility. The numerical simulations showed good approximation with the velocity measurements for two different geometries. To evaluate the accuracy of the numerical results, several experimental tests, using two different geometries, were firstly developed, identifying perturbations in the flow measurements, followed by analysis in a real case study.

2. Electromagnetic Flowmeters

Flowmeters are one of the most important and used devices to measure accurately, the volume flow rate [6]. Nowadays there are several types of flowmeters, but the most important for flow measurement are the electromagnetic and the ultrasonic ones. They are characterized for their high accuracy and self-monitoring [7]. According to the authors of [1], electromagnetic flowmeters are only disturbed by existing particles that may change the magnetic properties of each fluid. Properties like temperature, viscosity and the fluid density do not affect the measurements. However, as mentioned in [8], a steady regime is a necessary condition to guarantee accurate measurements.

These flowmeters have two different elements: the primary and the convertor (Figure 1). The first element corresponds to a hollow circular pipe with coils along its length and is set in the pipeline [9,10]. The flow passing through the section, creates an electromagnetic field which is proportional to the volume flow rate. The convertor is the brain element: it creates a magnetic field, reads the voltage, displays the data, and generates outputs. The convertor displays the volume flow rate and the amount of volume passed through.

(a)   (b)

Figure 1. Electromagnetic flowmeters components: (**a**) primary element; (**b**) convertor (adapted from [1]).

According to several manufactures, electromagnetic flowmeters assure an accuracy higher than 0.2%, as long as the flow velocities are higher than 1 m/s and the installation requirements are fulfilled. The accuracy of electromagnetic flowmeters depends on the velocity, according to Equation (1) [1] and as represented in Figure 2.

$$\begin{cases} -40U + 6\% & U < 0.1 \, m/s \\ 8(U - 0.55)^2 + 0.38\% & 0.1 \, \text{m/s} \leq U \leq 0.5 \, \text{m/s} \\ -0.4U + 0.6\% & 0.5 \, \text{m/s} \leq U < 1 \, \text{m/s} \\ 0.2\% & U \geq 1 \, m/s \end{cases} \tag{1}$$

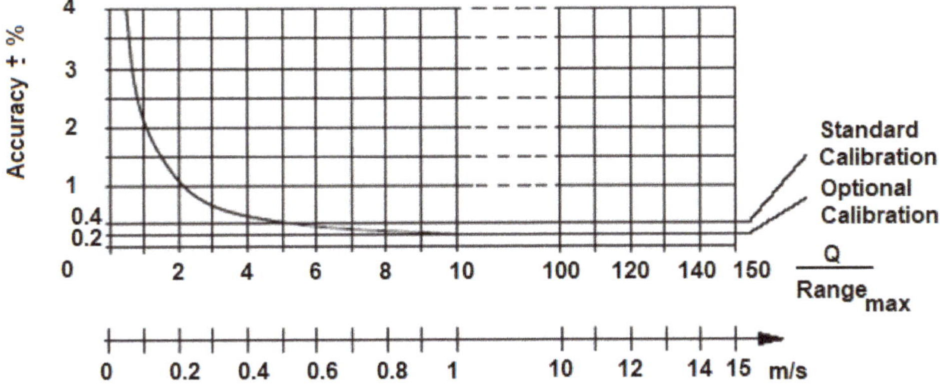

Figure 2. Electromagnetic flowmeter accuracy curve [1].

In order to measure the volume flow rate, the Faraday principle is applied in electromagnetic flowmeters. In 1831, Michael Faraday discovered that if an electric conductor is moving in a magnetic field, perpendicular to the direction of the motion, an electrical current is induced and proportional to the magnetic field force, as well as the velocity. If the conductor is water, the flow passing through a magnetic field induces an electrical current proportional to the flow velocity. Figure 3 represents the operating principle, which is important to understand the following developments.

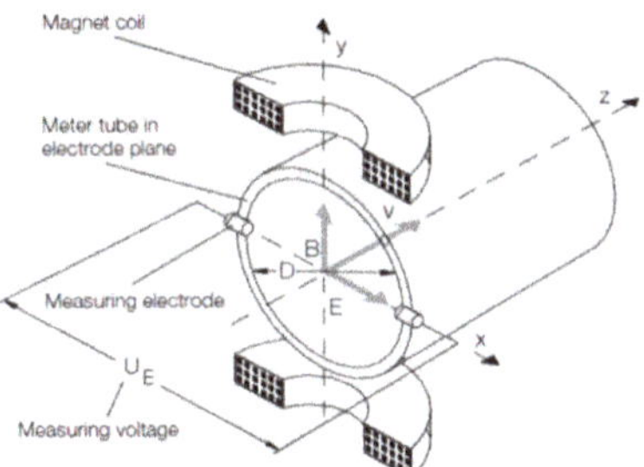

Figure 3. Operating principle of an electromagnetic flow meter [1].

Through mathematical manipulation, the electrical current, UE, induced by the flow passage is directly proportional to the value of the volume flow rate, Q, or $U_E \sim Q$. According to [1], the electrical current induced by the flow is taken into account only in the cross section defined by the electrodes, perpendicular to the flow. In other words, only the parallel component of the velocity is relevant for the volume flow rate measurement [11]. This means that the tri-dimensional nature of flow is disregarded.

The flow velocity profile is not the same throughout the entire cross section. For that reason, to prevent over or under-evaluations of the flow velocity, the suppliers of this equipment use a weighting factor, W. Figure 4 represents the weighting factor distribution in a cross section.

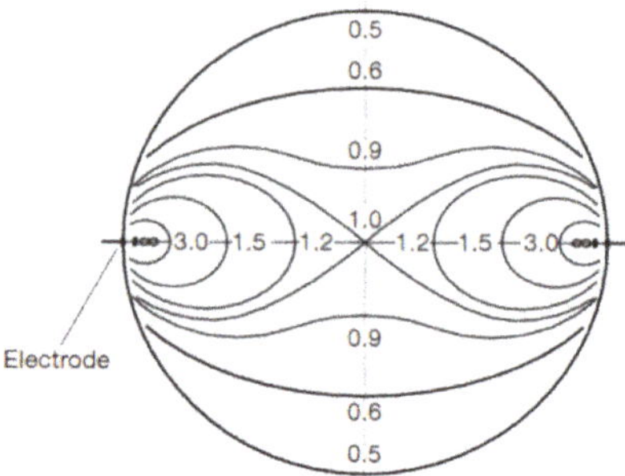

Figure 4. Weighting factor distribution *W* in the electrode plane [1].

In Figure 4, each point in the cross section has different weighting factors associated. The sum of the product between the velocity and the respective weighting factor corresponds to the electrical current, which is proportional to the volume flow rate. Although it is a good method to determine the volume flow rate in a homogenous constant magnetic field along symmetric velocity profiles, this formulation does not provide good results for non-symmetric velocity profiles. In these cases, it would over evaluate some values and under evaluate others [1], leading to a inaccurate volume flow rate. To avoid this, the suppliers of the equipment consider a magnetic induction field, *B*, inversely proportional to the weighting factor *W*, Equation (2):

$$W \times B = const \tag{2}$$

According to Equation (2), for a cross section region in which the weighting factor is small, the magnetic induction field is increased, and vice versa. This action ensures good results even for non-symmetrical velocity profiles [2].

3. Equipment and Layout

3.1. Flow Meter Installation

According to several authors [1,2,4], electromagnetic flowmeters (Figure 5) are only disturbed by the existence of particles that might change the magnetic properties of the fluid. Thus, it is necessary to guarantee the existence of linear flow paths, to insure the measurement accuracy. It is therefore necessary to meet the minimum installation requirements for each geometry, specified by each manufacture. Figure 5 presents the installation requirements imposed by the suppliers for the most common geometries.

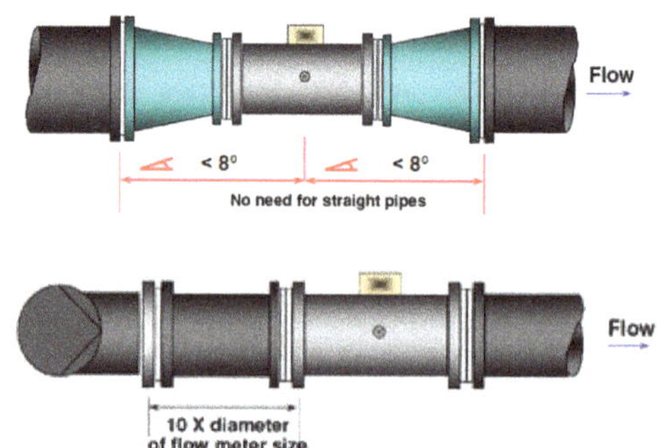

Figure 5. *Cont.*

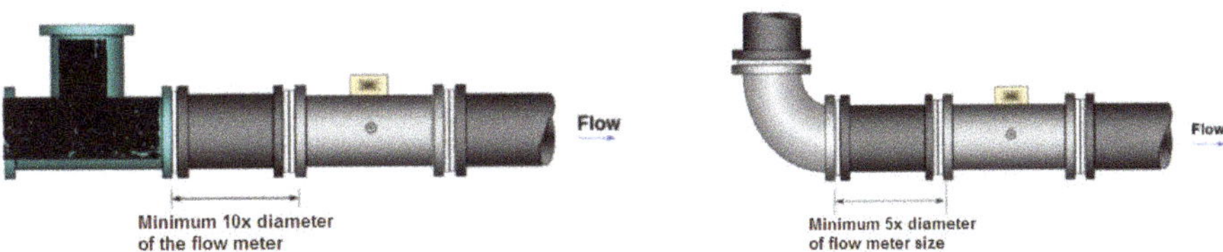

Figure 5. Installation requirements for different pipe layouts.

3.2. UDV Installation

The UDV operating principle is the MET-FLOW approach—an ultrasonic probe is placed near the pipe wall with a certain slope. The ultrasound is emitted and travels across the pipe cross section. When the ultrasound hits a fluid particle, some energy of the ultrasound disperses and produces an echo, which reaches the probe. Then, by a mathematical manipulation, the equipment delivers a velocity value. Figure 6 presents the UDV operating principle.

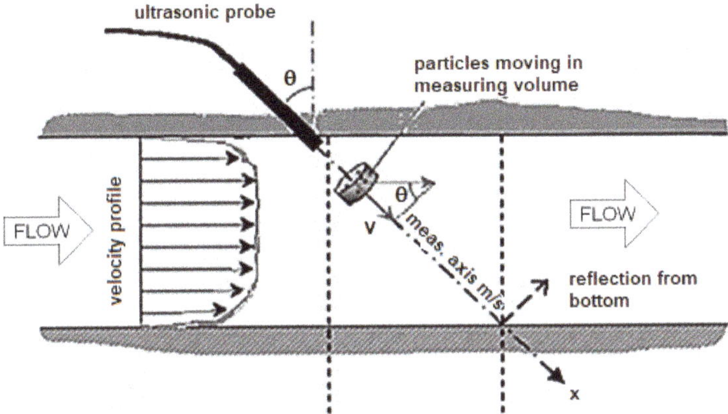

Figure 6. Ultrasonic Doppler velocimetry (UDV) operating principle (adapted from [12]).

The UDV uses an ultrasonic wave in order to provide the velocity profile. At a certain time (e.g., *t*1), a burst is emitted. This burst propagates inside the liquid. At time *t*2, the burst touches the particle. If the sizes of the particle are much smaller than the wave length, only a very small echo is generated (scattering effect). This echo goes back in direction to the transducer, while the main energy continues its propagation [11,13]. At time *t*3, the echo reaches the transducer. The depth of the particle, $Depth = C/2(t3 - t1)$, can be determined from the traveling time ($t3 - t1$), where C is the sound velocity of the acoustic wave in the liquid. Following each emission, the echo signal is sampled at a fixed delay after the emission. This delay defines the depth. However, it the particle moves between the successive emissions the sampled values taken at time *ts* will change over the time [14]. Depending of the shape of the emitted signal, these values may form a sinusoidal signal. The frequency, F_d, of this sinusoidal signal, which is named Doppler frequency, is directly connected to the velocity of the particle, which is given by the Doppler equation [14]:

$$V = \frac{F_d C}{2 F_e Cos\theta} \tag{3}$$

where F_e is the frequency of the emitted burst.

UDV offers instantaneously a complete velocity profile. Unfortunately, as the information is available only periodically, the maximum velocity (V_{max}) exists for each pulse repetition frequency (F_{prf}) [11]:

$$V_{max} = \frac{F_{prf}C}{4F_e Cos\theta} \tag{4}$$

Thus, the maximum measurable depth (P_{max}) is also defined by the pulsed repetition frequency Equation (5), and consequently the product of P_{max} and V_{max} is constant, and is given by Equation (6) [13,14]:

$$P_{max} = \frac{C}{2F_{prf}} \tag{5}$$

$$P_{max} \times V_{max} = \frac{C^2}{8F_e Cos\theta} \tag{6}$$

In the presence of air, the equipment signal is not able to read the signal, therefore, the probe needs to be well installed. Thus, the probe is set in a specific probe holder (Figure 6), which is a plastic rectangle with several holes, where each of it has a certain angle associated. This probe holder has two functions: to guarantee the stability to the probe, since it makes possible to attach it to the pipe; and also make sure that there is no air between the pipe and the probe. This is accomplished by inserting a gel in the hole where the probe will be installed, always ensuring contact with it.

Moreover, since UDV uses very sensitive equipment, specially to electromagnetic noise, its reading may be compromised by the electromagnetic field induced by the flowmeters.

3.3. Experimental Facility

In order to assess the computational results provided by the CFD model, intensive campaigns were developed in an experimental facility, schematically represented in Figure 7, and adapted to two continuous geometries and operating conditions. Regarding this pipe system layout, UVD measurements were made in section A (in Figure 7a). Two different volume flow rates, corresponding to 100 and 12 m^3/h were tested. In each UDV measurement, the velocity was captured in 100 different points in a total of 100 profiles.

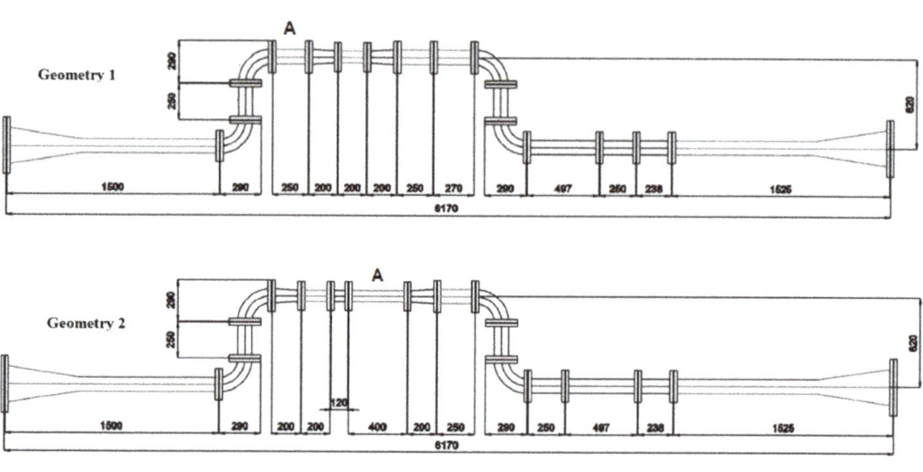

Figure 7. *Cont.*

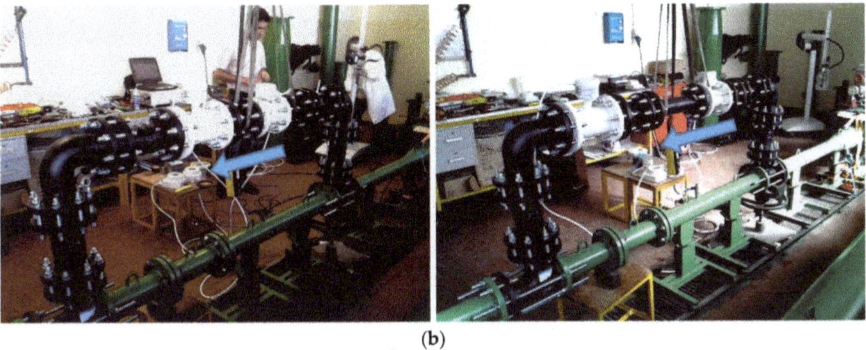

(b)

Figure 7. System layouts 1 and 2: (**a**) facility scheme; (**b**) lab installation, with flow direction identified by the blue arrow.

The results obtained are presented in Table 1 for section A (Figure 7). The flowmeter error decreases as the velocity increases, as it happens for $Q = 100$ m^3/h for both layouts.

Table 1. Test results: experiments and relative errors achieved in section A, for different volume flow rates.

Geometry	$Q_{theoretical}$ (m^3/h)	Tests Results					Error		
		V_{ND100} (L)	$V_{reference}$ (L)	$t_{theoretical}$ (s)	t_{real} (s)	$Q_{reference}$ (m^3/h)	ND100 (%)	Re (-)
1	100	4980	5000	180	173	104	−0.40%	365,631
	12	1006	1018	305	285	13	−1.18%	45,704
2	100	5026	5000	180	172	105	0.52%	369,147
	12	1035	1020	306	295	12	1.47%	42,188

The velocity profiles measured in section A, using the UDV, are presented in Figure 8 for 100 m^3/h and 12 m^3/h, respectively.

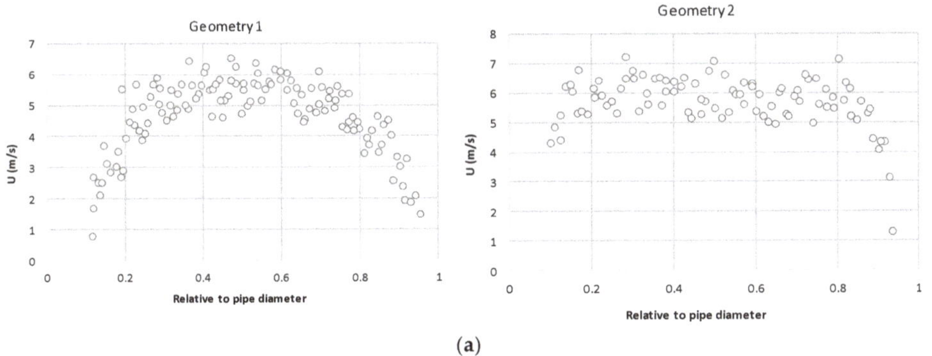

(a)

Figure 8. Cont.

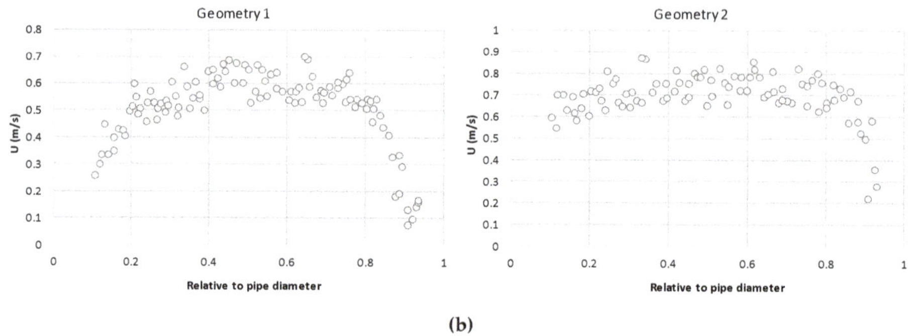

(b)

Figure 8. UDV profiles related to geometries 1 and 2: (**a**) for 100 m^3/h; (**b**) for 12 m^3/h.

4. CFD Model

4.1. Governing Equations

Computational fluid dynamics (CFD) provide a qualitative and quantitative prediction of fluid flows by means of mathematical and numerical methods. These simulation tools represent an important technological advance towards the detailed understanding of the flow, allowing theoretical considerations regarding the physical behavior of the flow, with mathematical formulations for tri-dimensional modelling analyses [7]. These models make possible, not only to study the behavior of turbulent and laminar flows, but also the multiple forms of exchanges of energy, flow phases, vorticity and turbulence levels [9].

The CFD model used in this work was COMSOL Multiphysics 4.3.b, which presents accurate results for several fluid flow problems [10]. COMSOL is a finite element method (FEM) software, which uses the mass conservation and the RANS (Reynolds averaged Navier–Stokes) equations as governing flow equations:

$$\frac{\partial \rho}{\partial t} + \frac{\partial (\rho u)}{\partial x} + \frac{\partial (\rho v)}{\partial y} + \frac{\partial (\rho w)}{\partial z} = 0 \tag{7}$$

$$\rho \bar{u}_j \frac{\partial \bar{u}_i}{\partial x_j} = \rho \bar{g}_i + \frac{\partial}{\partial x_j} \left[-\bar{p}\delta_{ij} + \mu \left(\frac{\partial \bar{u}_i}{\partial x_j} + \frac{\partial \bar{u}_j}{\partial x_i} \right) - \rho \overline{u'_i u'_j} \right] \tag{8}$$

FEM is a computational method that divides an object into smaller elements. Each element is assigned to a set of characteristic equations that are then solved as a set of simultaneous equations to estimate the behavior of the object [15]. From the available turbulence models, the k-ε model was selected. The k-ε models [16] are the most common and most used models worldwide mostly for industrial applications due to its good convergence rate and relatively low memory requirements. This model solves two variables: k, the turbulence kinetic energy and ε, the rate of dissipation of turbulence kinetic energy. This turbulence model relies on several assumptions, the most important of which is that the Reynolds number is high enough. It is also important that the turbulence is in equilibrium in boundary layers, which means that the production equals the dissipation. These assumptions limit the accuracy of the model because they are not always true, since it does not respond correctly to flows with adverse pressure gradients that can result in under predicting the spatial extension of recirculation zones [17]. Furthermore, in the description of rotating flows, the model often shows poor agreement with experimental data [18]. In most cases, the limited accuracy is a fair trade-off for computational resources saved compared to more complex turbulence models.

While it is possible to modify the k-ε model so that it describes the flow in wall regions, this is not always desirable because of the very high resolution requirements. Instead, analytical expressions are used to describe the flow at the walls [15]. These formulations are known as wall functions.

Wall functions ignore the flow field in the buffer region and analytically compute a nonzero fluid velocity at the wall [17–19].

Thus, by using wall functions, the wall lift-off in viscous units δ_w^+ needs to be checked. This value alerts if the mesh at the wall is fine enough and should be 11.06 everywhere. If the mesh resolution in the direction normal to the wall is too coarse, then this value will be greater than 11.06, and a finer boundary layer mesh needs to be applied in these regions. The second variable that should be checked when using wall functions is the wall lift-off δ_w (in length units) [18]. This variable is related to the assumed thickness of the viscous layer and should be small relative to the surrounding dimensions of the geometry. If it is not, then the mesh in these regions must be refined, as well.

The wall functions in COMSOL are such that the computational domain is assumed to start a distance δ_w from the wall (Figure 9).

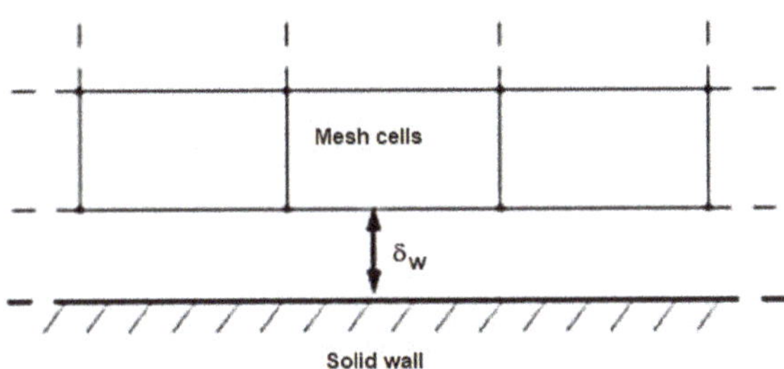

Figure 9. Computational domain starts a distance δ_w from the wall [13].

Nevertheless, in all simulations the fluid was water, with constant density and viscosity equal to 999.62 Kg/m^3 and 1.0097×10^{-6} m^2/s, respectively.

In the CFD model, three types of boundary conditions were assigned: inlet, outlet and solid walls. For the inlet boundary condition, the pressure was set, as for the outlet condition, the average velocity. The no-slip condition was considered, which stated that the walls were impermeable. Thus, the boundary conditions used to defined the inlet condition, are governed by the set of the following equations [16]:

$$p = p_0$$
$$\left[(\mu + \mu_T)\left(\nabla u + (\nabla u)^T\right) - \tfrac{2}{3}\rho k l \right] \times n = 0 \tag{9}$$
$$k = \tfrac{3}{2}\left(U_{ref} I_T\right)^2 \qquad \varepsilon = C_\mu^{3/4}\frac{k^{3/2}}{L_T}$$

where p_0 is the input value (of pressure), I_T is the turbulent intensity, L_T corresponds to the turbulence length scale, l is the mixing length defined by [20], and U_{ref} is the reference velocity scale.

For the outlet condition, the boundary conditions are expressed by Equation (10), where U_0 corresponds to the average velocity (input value) and, the first equation represents the normal outflow velocity magnitude:

$$u = U_0 n$$
$$\nabla k \times n = 0 \qquad \nabla \varepsilon \times n = 0 \tag{10}$$

4.2. Mesh Definition and Solution Convergence

As for the mesh definition, the geometry was discretized into smaller units, called mesh elements. Its resolution and element quality are important aspects to take into account, when validating the model, since the decreasing of resolution can originate low accuracy results [21]. Meanwhile, low mesh element quality can lead to convergence issues [22–24].

All calculations have been performed on a PC (Intel 5, CPU 3.90 GHz, RAM 8 GB) with 4 cores and threads running in parallel. The model default uses a physic controlled mesh, which defines, automatically, the size attributes and operations sequences necessary to create a mesh adapted to the problem. This mesh is automatically created and adapted for the model's physics settings. The default physics-controlled meshing sequences create meshes that consist of different element types and size features, which can be used as a starting point to add, move, disable, and delete meshing operations. Each meshing operation is built in the order it appears in the meshing sequence to produce the final mesh [23]. Customizing the meshing sequence helps to reduce memory requirements by controlling the number, type, and quality of elements, thereby creating an efficient and accurate simulation [25].

For the fluid-flow model, since the mesh is adapted to the physic setting, the mesh is finer than the default one, with a boundary layer (Figure 10a) in order to solve the thin layer near the solid walls where the gradients of the flow variables are high [26–29].

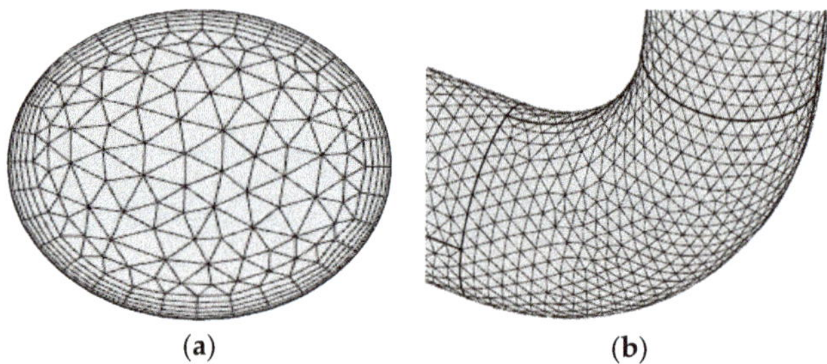

(a)   (b)

Figure 10. Physics-controlled mesh used: (**a**) Boundary layer mesh; (**b**) Refinement near a curve.

To ensure a proper and acceptable accuracy of the results, COMSOL uses an invariant form of the damped Newton method. Starting with Z_0, the linear model (MUMPS) is solved for the Newton step (δZ) [16]. Afterwards, a new iteration is calculated, according to Equation (7), where λ' is the damping factor:

$$Z_1 = Z_0 + \lambda' \delta Z$$
$$|\lambda'| < 1 \tag{11}$$

The model estimates the error of the new iteration and, if the error of the current iteration is bigger than the previous one, the code decreases the damping factor and a new iteration process restarts. This procedure will occur until either the error is smaller than the error calculated in the previous iteration or the damping factor reaches its minimum value (i.e., 1×10^{-4}). When a successful step is reached, the algorithm computes the next iteration. The iteration process finish when the relative tolerance exceeds the relative computed error. The model stops the iteration when the relative error is smaller than 1×10^{-3} and the damping factor is equal to 1. Otherwise the solution would not converge and the iteration would continue. Figure 11 presents the convergence solution reached.

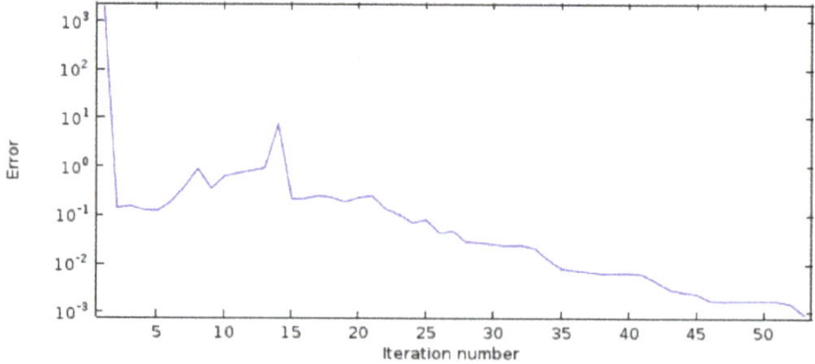

Figure 11. Convergence solution.

4.3. Calibration and Validation

According to the input data and the respective geometries, presented in Table 2 and in Figure 12, numerical simulations were made and compared with the experimental tests.

Table 2. Boundary, mesh and study conditions.

Characteristics	100 m^3/h	12 m^3/h
Inlet	5.6 bar	5.8 bar
Outlet	0.59 m/s	0.07 m/s
Wall	No-slip	
Mesh	Physics-controlled	
Flow conditions	Steady state	

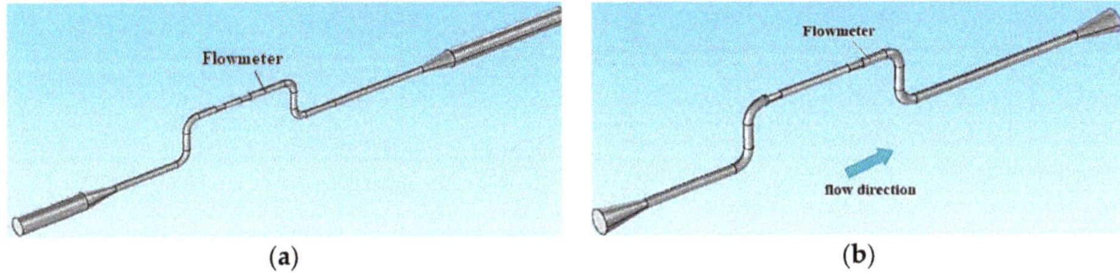

(a) (b)

Figure 12. Model configuration: (**a**) geometry 1; (**b**) geometry 2.

Figure 13 presents the velocity profiles obtained in the CFD model and in the experimental tests, for the same section of the UDV measurements.

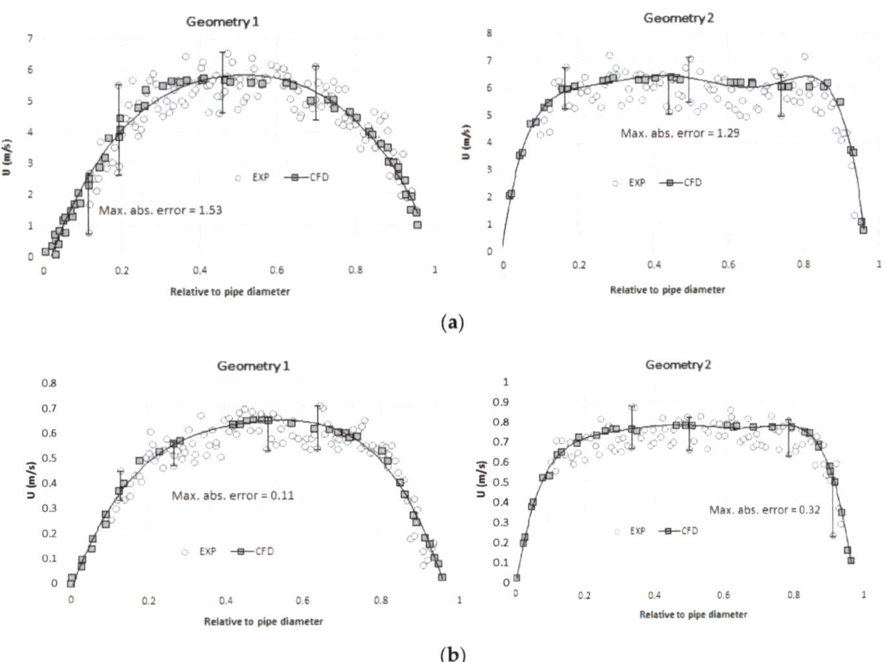

Figure 13. Comparison of velocity distribution profiles obtained through CFD model and experiments: (**a**) for 100 m^3/h; (**b**) for 12 m^3/h.

The velocity profiles in Figure 13, related to UDV, presents some spikes. This behavior is justified by the presence of fluctuations in the velocity and the existence of vortexes, both characteristic of deviations on the flow direction. Despite this, the results present a good approximation between the experimental data and the CFD results. In Figure 14 is represented the velocity contours in the pipe cross section for the two different volume flow rates.

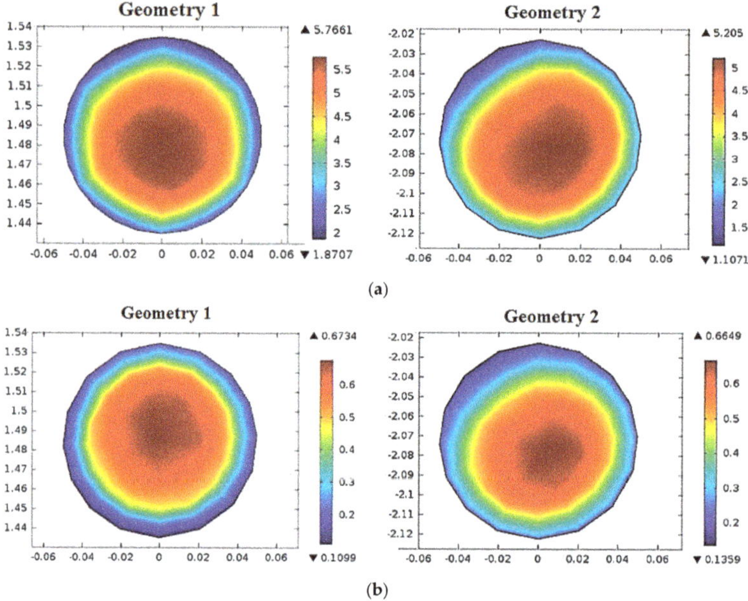

Figure 14. Velocity contours in the pipe cross section: (**a**) for 100 m^3/h; (**b**) for 12 m^3/h.

In order to assess the associated errors of the volume flow rate measured by the flowmeter, if the velocity distribution is equal to the numerical model, several automatic procedures were implemented. The first step involves the definition of limits presented in Figure 4. The relevant data for this analysis was the one associated to the electrodes cross section of the flowmeter. The data redrawn from the model is a plan with the three coordinates, x, y and z and the average velocity, U. Each two coordinates are associated a velocity and each limit to a certain weight. Therefore, the velocity at points located near the limits (with a maximum error of 0.5%) was multiplied by the corresponding weighting factor (Figure 15).

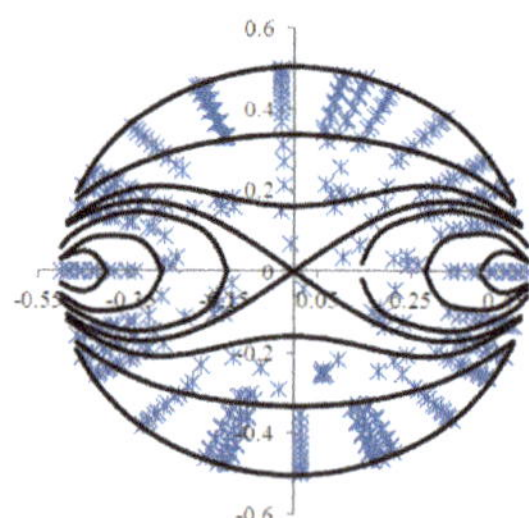

Figure 15. Weighting function distribution of uniform magnetic field point electrode electromagnetic flowmeter. Calculated limits (black continuous lines) analogous to Figure 4. Data provided by the model represented by the blue ticks.

The subsequent step is the interpolation of points between limits, which were multiplied by the correspondent weighting factor. This procedure was made for all limits except the ones that are closer to the electrodes. The limits near the electrodes were very difficult to assess, since closely to the electrodes the weight was not entirely known. Thus, another assumption was taken, i.e., the weighting factor was calculated according to Figure 16. This function was defined and validated through the available experimental data. The x variable is the value resulted from the subtraction of the number of the total data point with the ones in the region near to the electrodes section.

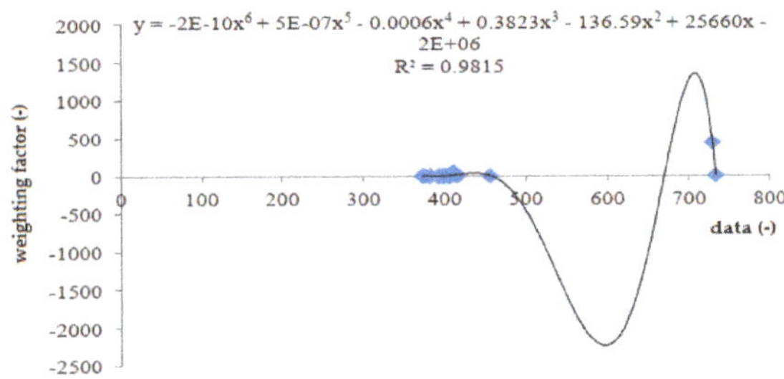

Figure 16. The best function for the calculus of the factor for the region near the electrodes to fit the experiments (blue marks represent the experimental data used).

Knowing the remaining factor, the velocity was obtained through the sum of the velocity of each point, multiplied by the corresponding weighting factor, and then divided by the sum of the weighting factors. The error was then determined (Table 3) by applying Equation (12).

$$Error[\%] = \frac{U_{calculated} - U_{boundary\ conditon}}{U_{boundary\ condition}} \times 100 \tag{12}$$

The theoretical error of the flowmeters ought to increase with the decrease in the volume flow rate. Thus, it is clear that the error associated to 12 m³/h are bigger than 100 m³/h (Table 3). For the volume flow rate of 100 m³/h it is verified that the error associated to geometry 1 is the smallest one. From the two experimental tests, geometry 2 corresponds to the worst scenario, since the pipe is not long enough to dissipate the flow perturbations caused by the profile vertical curves.

The difference between CFD and the lab tests is twofold: minor installation problems, such as the position of the gasket inserted in between flanges causing obstructions to the reading, and the level of detail of the CFD model.

Table 3. Summary of the errors calculated with the volumetric method for both geometries.

Geometry	Q = 100 m³/h		Q = 12 m³/h	
	Error		**Error**	
	Experimental	Model	Experimental	Model
1	−0.40%	−0.48%	−1.18%	−1.27%
2	0.52%	1.11%	1.47%	1.61%

The error associated to the position of the gasket is important and, in several situations, avoidable errors. In these experiments it is mathematically improbable to assume that the errors could be preventable. However, in practice, since the number of gaskets is much smaller, these errors can be disregarded if engineering good practices are follow.

5. Case Study

5.1. Geometry Layout

The pumping station layout is presented in Figure 17, where almost 90% of the total water volume is pumped. The flow arrives by the drive pipe, i.e., right to left side of Figure 17c, and follows through the pump (represented by the green section) by a vertical pipe in the top of the drive pipe. Afterwards, the flow reaches the delivery pipe with a 30° angle, sidewise, after passing two 90° curves.

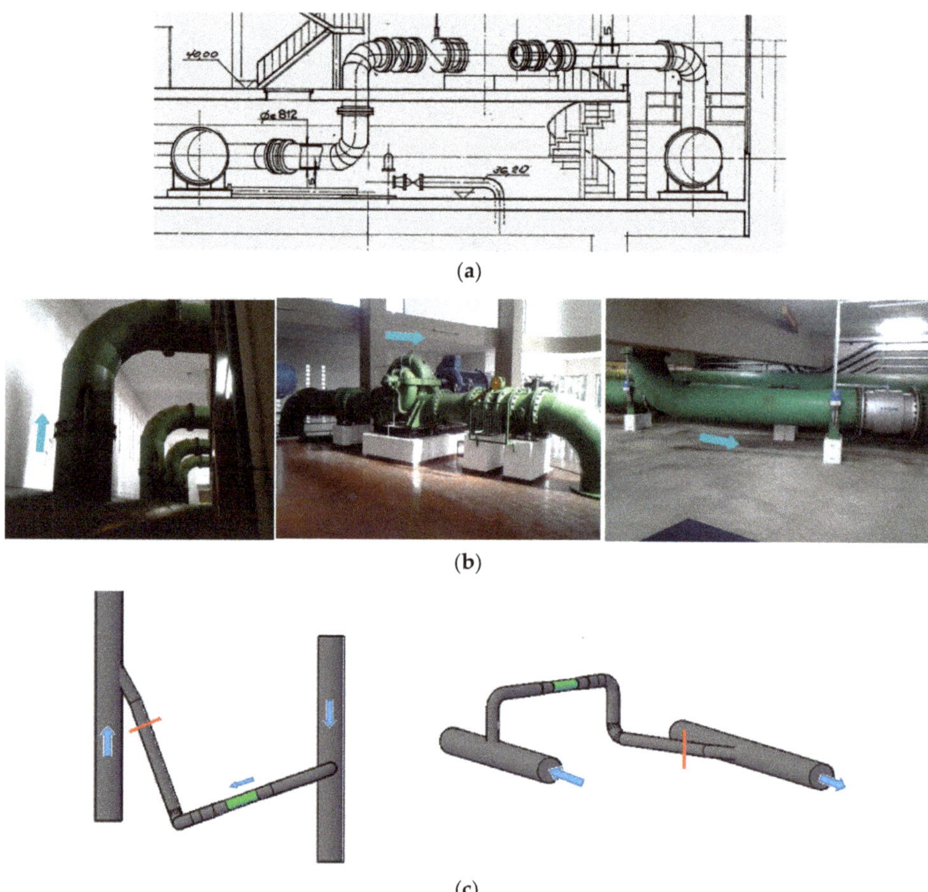

Figure 17. Pumping station layout (**a**); photographs of the different parts of the hydraulic system (**b**); modelling geometry—pump location identified by the green pipe section, flowmeter identified by the red line, flow direction identified by the blue arrow (plan view on the left and 3D view on the right) (**c**).

5.2. Simulations

The problems detected in situ were assumed to be correlated to the flowmeters in the pumping station. As the flow passes through the pump, the disturbances from upstream can be disregarded, i.e., considering only the perturbations caused by the pump. Thus, since the pressure is measured just downstream of the pump, the geometry used for simulating the disturbances is presented in Figure 18 and the related characteristics of the pipe and the flowmeter are presented in Table 4.

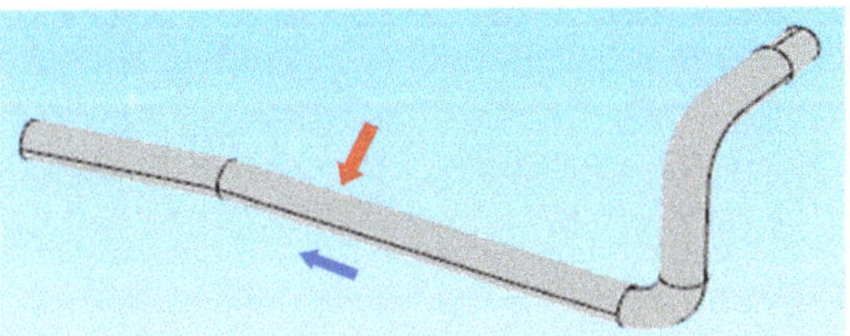

Figure 18. System layout modelling geometry—with the flowmeter section identified by the red arrow and flow direction identified by the blue arrow.

Table 4. Characteristic of the main pipe and flowmeter.

Material	Steel
Expansion	ND700 to ND800
Remaining pipes	ND800
Flowmeter	ND800

The simulations were performed by first defining the geometry, after which was necessary to identify the characteristics of the fluid (i.e., water) and the boundary conditions. Lastly the mesh definitions were chosen. The boundary conditions and other important features are described in Table 5.

Table 5. System layout existing situation: simulation input values and characteristics.

Inlet	9.5 bar
Outlet	1.5 m/s
Wall	No-slip
Mesh	Physics-controlled
Flow conditions	Steady state

Figure 19 presents the streamlines velocities and the velocity distribution in the cross section of the flowmeter, respectively. In the flowmeter section, the streamlines appear to be parallel to each other.

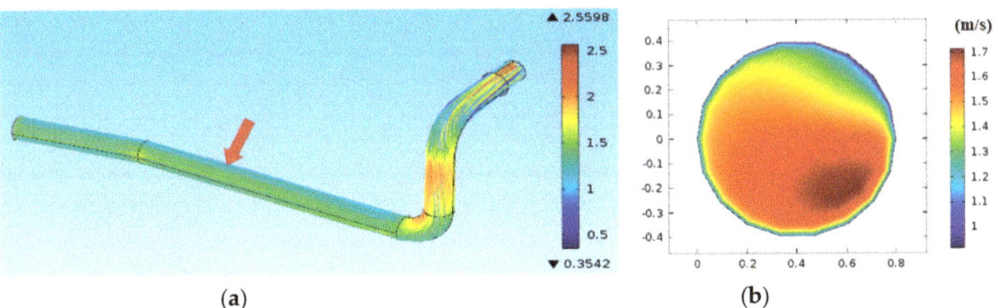

(**a**)   (**b**)

Figure 19. Streamlines simulation along the hydraulic circuit for the pipe branch system layout (flowmeter section identified by the red arrow), in m/s (**a**); Velocity distribution in the cross section of the flow meter (**b**).

The velocity changes largely within the cross section, from zero near the wall to over 1.7 m/s in the right lower side, where the velocity attains the highest value. This behavior was not expected because the flow passes through two 90° curves after the pump, therefore, it would be reasonable to anticipate that the perturbations induced would be, in practice, neglected, according to manufactures experience. Nevertheless, it is important not to forget that the second curve presents a rotation in axis z. Therefore, the non-symmetric velocity is due to the existing geometry (Figure 19b).

For each point, provided by the numerical model, the correspondent weighting factor was estimated. The velocity measured by the flowmeter was calculated dividing the product of the velocity with the weighting factors by the sum of all the weighting factors. The error associated to the flowmeter was about 0.71%, determined according to Equation (8). The error calculated by the manufacture, through water balances, was close to −1%. Since the errors are alike, the model is considered to be validated and the results accurate.

5.3. Solution

In order to achieve a solution that presents a lower error, 3 different procedures can be adopted: (1) changing the geometry but maintaining the values of volume flow rate and pressure; (2) changing the inlet and outlet conditions but maintaining the geometry; and (3) using both solutions. The chosen procedure was the first one, since the volume flow rate and pressure demanded downstream remain the same.

The previous results showed that the existing layout would not be enough to guarantee the dissipation effects and the flow uniformity. For that reason, a new pipe with 20 m of length was added in order to make possible the complete dissipation of disturbances. Considering the flowmeter in the same location, as the previous simulation (i.e., the flowmeter located between 24 m of straight pipe upstream and 4 m downstream), the new proposed geometry corresponds to Figure 20.

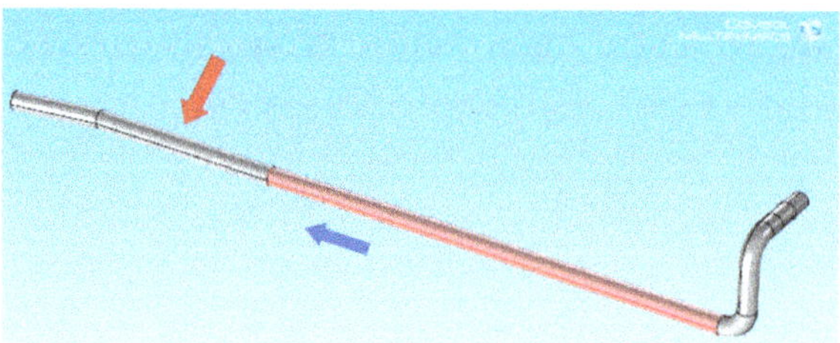

Figure 20. System layout proposed modelling geometry—the flowmeter section represented by the red arrow; flow direction identified by the blue arrow (added pipe in red).

For the same boundary conditions and mesh features, presented in Table 5, the streamlines obtained are presented in Figure 21a as well as the velocity distribution across the electrodes plan (Figure 21b).

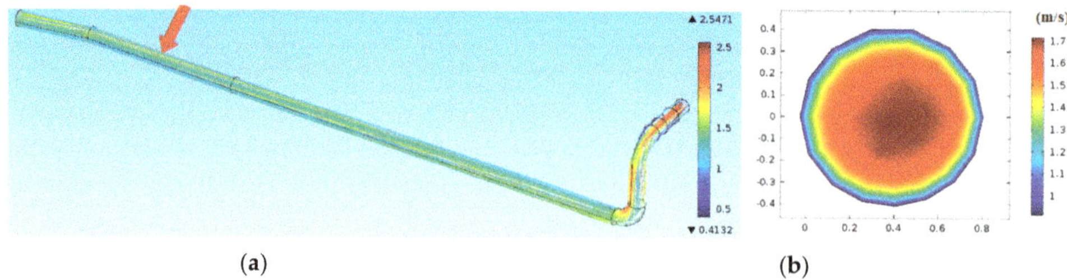

(a)   (b)

Figure 21. Streamlines simulation along the hydraulic circuit for the proposed system layout (flowmeter section identified by the red arrow), in m/s (**a**); velocity distribution in the electrodes cross section for the proposed geometry (**b**).

This profile presents minor perturbations compared to the one in Figure 18b. However, the velocity distribution is not exactly symmetric. The perturbation is still noticeable several meters ahead of the last singularity. For this amount of volume flow rate and pressure, the error associated to this simulation is 0.37%, which is closer to the reference value that the flowmeters suppliers assure for a flow velocity higher than 1 m/s (i.e., 0.2%).

6. Conclusions

The measurement problem has a significant influence in the correct management of technical, hydraulic and economic efficiency of water companies. Since the hydraulic circuits have, normally, two flowmeters, a water balance is possible. These balances are important tools to detect leaks throughout the supply and distribution processes. If the measurement is not accurate enough, the balances do not match and, subsequently, these tools loss their importance.

The procedure used to analyze the associated errors, revealed to be enough showing good results for the experiments and the case study. Nevertheless, more scenarios need to be studied in order to assess if the procedure developed can be applied for every case.

Regarding the simulation results, the model presented an error very similar to the one verified in the experimental tests developed by [30]. According to the results achieved, the errors calculated are very significant for an equipment with high accuracy. Consequentially, errors associated to the installation and the geometry are very important with relevant issues that are not always taken into account by engineers, responsible for the design, nor by teams responsible for the installation of this equipment. If these factors are not addressed properly, the flow measurement is not accurate.

Regarding the proposed solution, the geometry to minimize the installation errors is not feasible, since the facility is already built according to manufacturer's conditions. This could be overcome if well-defined automatic tools were applied to estimate the uncertainty of flow measurements. Using CFD models, a faster error approximation from a certain type of geometry, volume flow rate, and pressure could be obtained.

Moreover, this research intends to assess that the installation requirements proposed by the manufactures are not sufficient to dissipate such uncertainties. Thus, it is reasonable to emphasize the importance of flow measurement for water companies, regarding more efficient and rational water use and management.

Author Contributions: The author H.M.R. has contributed with the idea, in the revision of the document and in supervising the whole research. M.S. and M.B. developed the experimental tests and the CFD modelling and the analysis of the results between CFD and experiments. H.M.R. and A.C. revised the final document.

Funding: This research was funded by REDAWN, EAPA_198/2016.

Acknowledgments: The authors wish to thank to the project REDAWN (Reducing Energy Dependency in Atlantic Area Water Networks) EAPA_198/2016 from INTERREG ATLANTIC AREA PROGRAMME 2014–2020 and CERIS-IST and the Hydraulic Laboratory, for the support in the conceptual developments and experiments.

Conflicts of Interest: The authors declare no conflict of interest.

References

1. Frenzel, F.; Grothey, H.; Habersetzer, C.; Hiatt, M.; Hogrefe, W.; Kirchner, M.; Lütkepohl, G.; Marchewka, W.; Mecke, U.; Ohm, M.; et al. *Industrial Flow Measurement Basics and Practice*; ABB Automation Products GmbH: Gottingen, Germany, 2011.
2. Sheng, H.; Lihui, P.; Nakazato, H. Computational fluid dynamics based sound path optimization for ultrasonic flow meter. *Chin. J. Sci. Instrum.* **2009**, *30*, 852–856.
3. Cardoso, A.H. *Hidráulica Geral I—Apontamentos Complementares das Aulas Teóricas*; IST: Lisboa, Portugal, 2009.
4. Lysak, P.D.; Jenkins, D.M.; Capone, D.E.; Brown, W.L. Analytical model of an ultrasonic cross-correlation flow meter, part 1: Stochastic modeling of turbulence. *Flow Meas. Instrum.* **2008**, *19*, 41–46. [CrossRef]
5. Pope, S.B. *Turbulent Flows*; Cambridge University Press: Cambridge, UK, 2000.
6. Quintela, A.C. *Hidráulica*; Fundação Calouste Gulbenkian: Lisboa, Portugal, 2002.

7. Matas, R.; Cibera, V.; Syka, T. Modelling of flow in pipes and ultrasonic flowmeter bodies. *EPJ Web Conf.* **2014**, *67*, 02073. [CrossRef]

8. EN 14154-2. *Water Meters—Part 2: Installation and Conditions of Use*; NSAI: Dublin, Ireland, 2005.

9. EN 14154-3. *Water Meters—Part 3: Test Methods and Equipment*; NSAI: Dublin, Ireland, 2005.

10. Fletcher, C.A.J. *Computational Techniques for Fluid Dynamics*; Springer: Berlin, Germany, 1991.

11. Brunone, B.; Berni, A. Wall shear stress in transient turbulent pipe flow by local velocity measurement. *J. Hydraul. Eng. ASCE* **2010**, *136*, 716–726. [CrossRef]

12. UVP Monitor –User's Guide" Model UVP-DUO with Software Version 3, Met-Flow, 2014.

13. Hammoudia, M.; Legrand, J.; Si-Ahmeda, E.K.; Salem, A. Flow analysis by pulsed ultrasonic velocimetry technique in Sulzer SMX static mixer. *Chem. Eng. J.* **2008**, *139*, 562–574. [CrossRef]

14. Haavisto, S.; Syrjnen, J.; Koponen, A.; Mannien, M. UDV Measurements and CFD Simulation of Two-Phase Flow in a Stirred Vessel. In Proceedings of the 6th International Conference on CFD in Oil & Gas, Metallurgical and Process Industries, Trondheim, Norway, 10–12 June 2008.

15. Georgescu, A.; Bernad, S.; Georgescu, S.-C.; CoSoiu, I. COMSOL Multiphysics versus FLUENT: 2D numerical simulation of the stationary flow arround a blade of the achard turbine. In Proceedings of the 3rd Workshop on Vortex Dominated Flows, Timisoara, Romania, 1–2 June 2007.

16. COMSOL 4.3. *COMSOL Multiphysics Reference Guide*; COMSOL AB: Stockholm, Sweden, 2012.

17. Speziale, C.G. On nonlinear k-l and k-ε models of turbulence. *J. Fluid Mech.* **1987**, *178*, 459–475. [CrossRef]

18. Simão, M. Fluid-Structure Interaction in Pressurized Systems. Ph.D. Thesis, Insituto Superior Técnico, Universidade de Lisboa, Lisboa, Portugal, 2017.

19. Gresho, P.M. Incompressible Fluid Dynamics: Some Fundamental Formulation Issues. *Ann. Rev. Fluid Mech.* **1991**, *23*, 413–453. [CrossRef]

20. Prandtl, L. *Guide à Traves de la Méchanique dês Fluides*; Dunod: Paris, France, 1952.

21. Lumley, J.L. Computational modelling of turbulent flows. *Adv. Appl. Mech.* **1978**, *18*, 123–176.

22. Lumley, J.L. Some comments on turbulence. *Phys. Fluids A* **1992**, *4*, 203–211. [CrossRef]

23. COMSOL 4.3. *COMSOL Multiphysics User's Guide*; COMSOL AB: Stockholm, Sweden, 2012.

24. Bakewell, H.P.; Lumley, J.L. Viscous sublayer and adjacent wall region in the turbulent pipe flow. *Phys. Fluids* **1967**, *10*, 1880–1889. [CrossRef]

25. Versteeg, H.K.; Malalasekera, W. *An Introduction to Computational Fluid Dynamics: The Finite Volume Method*; Pearson Education Limited: London, UK, 2007.

26. Çengel, Y.A.; Cimbala, J.M. *Fluid Mechanics—Fundamentals and Applications*; McGraw-Hill: New York, NY, USA, 2006.

27. Batchelor, G.K. *An Introduction to Fluid Dynamics*; Cambridge University Press: Cambridge, UK, 1967.

28. Abbott, M.B.; Basco, D.R. *Computational Fluid Dynamics—An Introduction of Engineers*; Longman Scientific & Technical: Harlow, UK, 1989.

29. Wosnik, M.; Castillo, L.; George, W.K. A theory for turbulent pipe and channel flows. *J. Fluid Mech.* **2000**, *421*, 115–145. [CrossRef]

30. Braga, F.; Fernandes, M. *Analise de Eventuais Perdas na rede de Adução da EPAL: Casos de Estudo*; EPAL: Lisboa, Portugal, 2009.

© 2018 by the authors. Licensee MDPI, Basel, Switzerland. This article is an open access article distributed under the terms and conditions of the Creative Commons Attribution (CC BY) license (http://creativecommons.org/licenses/by/4.0/).

Article

A Model for Selecting the Most Cost-Effective Pressure Control Device for More Sustainable Water Supply Networks

Irene Fernández García *, Daniele Novara and Aonghus Mc Nabola

Department of Civil Structural & Environmental Engineering, Trinity College of Dublin, Dublin D02 PN40, Ireland; novarad@tcd.ie (D.N.); amcnabol@tcd.ie (A.M.N.)
* Correspondence: g52fegai@uco.es

Received: 14 May 2019; Accepted: 19 June 2019; Published: 21 June 2019

Abstract: Pressure Reducing Valves (PRV) have been widely used as a device to control pressure at nodes in water distribution networks and thus reduce leakages. However, an energy dissipation takes place during PRV operation. Thus, micro-hydropower turbines and, more precisely, Pump As Turbines (PAT) could be used as both leakage control and energy generating devices, thus contributing to a more sustainable water supply network. Studies providing clear guidelines for the determination of the most cost-effective device (PRV or PAT) analysing a wide database and considering all the costs involved, the water saving and the eventual power generation, have not been carried out to date. A model to determine the most cost-effective device has been developed, taking into account the Net Present Value (NPV). The model has been applied to two case studies: A database with 156 PRVs sites located in the UK; and a rural water supply network in Ireland with three PRVs. The application of the model showed that although the investment cost associated to the PRV installation is lower in the majority of cases, the NPV over the lifespan of the PAT is higher than the NPV associated with the PRV operation. Furthermore, the ratio between the NPV and the water saved over the lifespan of the PAT/PRV also offered higher values (from 6% to 29%) for the PAT installation, making PATs a more cost-effective and more sustainable means of pressure control in water distribution networks. Finally, the development of less expensive turbines and/or PATs adapted to work under different flow-head conditions will tip the balance toward the installation of these devices even further.

Keywords: pumps as turbines; pressure reducing valves; energy recovery; leakage reduction; water-energy nexus

---

1. Introduction

World water use has risen considerably in recent decades and this growing tendency is likely to continue due to the expected population growth in the coming years [1]. In addition, large cities are increasingly concentrating a higher number of people, leading to water scarcity conditions in many urban areas [2]. One quarter of the largest cities in the world were recently estimated to be water stressed [3]. In Europe, and more specifically in its southern regions, a 24% decrease in renewable water resources has been detected during the period 1960–2010 [4]. Moreover, world energy consumption is also expected to increase by 28% in 2040 with respect to 2015 [5]. Therefore, the sustainable use of water and energy resources is currently one of the main concerns worldwide.

Water abstraction and distribution are among the activities in which the water-energy nexus plays an important role. In the European Union, as an example, 8% of the total energy consumption in 2014 was related to water supply [6]. In addition, it is estimated that 32 billion cubic meters per year (66% of the treated water) are lost in the water distribution process globally [7]. This is mainly due to the ageing infrastructure, the non-optimal design of the water supply systems, and the increase in water stress in urban areas [8]. Different strategies have been proposed to reduce energy consumption from fossil fuels in the water industry sector, by using renewable energy sources [9] or by recovering the excess of heat at the wastewater treatment plant [10], to name a few.

Leakage management can also play a major role in the reduction of energy consumption in the water sector. Thus, pressure control in water distribution networks is one of the most effective measures to reduce leakages because of the direct relation between pressure and leakage rate [11,12]. Hence, Pressure Reducing Valves (PRVs) play an important role in reducing water losses. The determination of the best location and the number of PRVs to install in a network has been widely analysed as an effective measure to decrease leakages. Gupta et al. [11] applied an optimization algorithm to determine the optimal PRVs location in a network, resulting in a leakage reduction of 21%. Fontana et al. [13] used real time control to manage the pressure in a PRV placed in a network inlet. This enabled optimal operation conditions by reducing pressure at nodes and therefore, leakages.

PRVs are a convenient device for reducing leakage. However, the energy dissipation that takes place in a PRV is wasteful of energy resources, and this energy could be recovered by substituting the PRV with a hydropower turbine. Thus, in addition to the reduction of water losses, a certain part of the energy in the network could be recovered, reducing greenhouse gas emissions and making the water supply system more sustainable [14]. Due to the potential of hydropower in these systems, several investigations have focused on the evaluation of the installation of these devices in water networks, showing that up to 40% of the gross power potential available in a PRV could be recovered by replacing the PRV with a PAT [15]. Other potential locations for installing hydropower turbines, such as break pressure tanks, water storages or water treatment plants, were also evaluated, estimating significant energy generation potential in some locations [16].

The main disadvantage historically associated with hydropower installations in water distribution networks was their lack of cost-effectiveness when the power output capacity is small [17]. The majority of power output capacities in typical PRVs in water distribution networks has been shown to lie in the range 1 to 15 kW [18]. However, recent developments in new micro-turbines which can operate under low flow/head conditions or new in-line Banki turbines with mobile regulating flap which enable the adaptation of the characteristic curve according to flow or pressure with reasonable high efficiencies may alter the results of these historical assessments which considered traditional turbine types [19–22].

A particular class of micro-turbines consists of pumps working in reverse mode, i.e., Pump As Turbines (PAT). These are devices that can be installed along distribution pipes to reduce pressure at nodes and recover energy [23–25], with significantly reduced investment costs compared to traditional turbines [26]. Optimization techniques have focused on identifying the best PAT location within water distribution networks to reduce leakage and generate energy. Giugni et al. [27] proposed a methodology to determine the optimal PAT location in water distribution networks considering two approaches: The reduction of leakage, and the maximization of energy. The authors demonstrated that the second approach involved higher energy production without decreasing pressure at nodes significantly. Other methodologies have considered the joint optimization of energy generation by means of PATs, leakage reduction and pump operation in order to reduce the energy costs associated with the water network [25,28]. The use of PATs as energy recovery devices within irrigation water distribution networks has also been proved feasible, with calculated annual energy recovery of up to 58 MWh and 271 MWh in selected networks [29,30].

The replacement of PRVs with PATs as a measure to generate energy and reduce leakage represents an efficient strategy to improve the sustainability of water supply networks. However, the sometimes small amounts of energy generated by a PAT can be interpreted as not achieving economic viability in some cases, making PRVs more attractive. For example, Fecarotta et al. [25] proposed an optimization procedure to determine the best location and the number of PATs considering both the maximization of the net present value (NPV) and the reduction of leakage. They included the water cost saving by leakage reduction in the estimation of the NPV and the algorithm provided a solution with 16 PATs to install on a selected network. However, 10 out of the 16 PATs were considered not economically feasible due to the low power generated. Therefore the authors proposed the installation of PRVs instead of PATs for those sites.

Some investigations have evaluated the replacement of PRVs by PATs as a measure to control pressure and generate energy, also analysing the most convenient element from an economic point of view [24,31,32]. However a comprehensive study giving guidelines for the installation of pressure reducing devices (PRV or PAT) for pressure reduction at specific nodes taking into account all the cost involved in both, and the eventual power generation have not been provided to date.

Therefore in this work a model has been developed to aid in choice between PAT and PRV in water distribution networks. The model focuses on the determination of the Net Present Value (NPV) taking into account the total cost associated with the installation of a PRV and a PAT as well as the cost savings associated with the reduction of leakage volume and the energy savings related to the PAT performance. The methodology has been applied to two case studies: The first one included a database with 156 PRVs sites located in Wales and West Midland regions of the UK; while the second case study included a water distribution network with three PRVs currently operating, located in a rural area in Ireland.

2. Methodology

2.1. Problem Approach

The model proposed in this work took into account all costs involved in the installation of a PRV and a PAT, as well as the sum of the incomes associated with the operation of both devices during their lifespan. Thus, NPV (€), which was previously used to determine the economic feasibility of the installation of PATs [25], was selected in this work to compare both devices:

$$NPV = -TC + \sum_{t=1}^{L} \frac{CS_t}{(1+r)^t} \tag{1}$$

where TC (€) is the total installation cost, t is an index related to year, L is the lifespan of the device considered, PRV or PAT (assumed as 15 years for both elements in this work). CS is the total cost saving at year t (€), and is determined by adding up the water cost saving, WCS, associated with the reduction of leakage volume after installing these devices and the energy cost saving, ECS, obtained by the PAT performance. r is the discount rate. A value of 0.05 for r was assumed in this work [33]. No management costs have been considered since the same cost has been assumed for both installation types.

The total installation cost was determined using Equation (2):

$$TC = C_{PRD} + C_{HE} + IC \tag{2}$$

where C_{PRD} is the cost of the pressure reducing device, PRD (PRV or PAT), C_{HE} the cost of the hydraulic elements required in the installation of each pressure reducing device, and IC is the installation cost. The determination of each term in Equation (2) is described in the following sections.

The estimation of WCS was carried out by the calculation of the water saving derived from the installation of a PRV or a PAT. When any of these devices is installed in a water supply network, pressure at nodes decreases and hence the leakage flow is reduced.

When the leakage flow at each pipe is known, this value can be assigned to each node and the discharge coefficient, c_i, which relates the leakage flow to pressure can be determined according to Equation (3):

$$ql_i = c_i \cdot P_i^{\beta} \tag{3}$$

where ql_i is the leakage flow at node i ($l\,s^{-1}$), β is the emitter exponent which takes into account the pipe material and the shape of the orifice (1.18 has been assumed in this work [34]) and P_i is the pressure (m) at node i. The coefficient c_i was determined by Equation (4):

$$c_i = \alpha \cdot 0.5 \cdot \sum_{j=1}^{Kji} L_{ji} \tag{4}$$

α is a coefficient ($l\,s^{-1}\,m^{-1-\beta}$), j is an index related to pipe, Kji is the number of pipes connected to node i, and L_{ji} is the length of the pipe j connected to node i (m). The coefficient α was determined using EPANET [35] which enabled the pressure dependent demand analysis. Once the coefficient α was determined, the effects of the performance of a PRD were evaluated in EPANET and WCS was estimated by:

$$WCS = c_w \cdot (Vl_0 - Vl_{PRD}) \cdot ndays \tag{5}$$

where c_w is the water cost ($0.3 \, € \, m^{-3}$ was assumed in this work [25]) and Vl_0 and Vl_{PRD} ($m^3\,day^{-1}$) are the leakage volume in the current situation and after installing the pressure reducing device, respectively. *ndays* was the number of days of the year. The leakage volume was obtained by applying Equation (6):

$$Vl = \frac{1}{1000} \cdot \sum_{h=1}^{24} Ql_h \cdot \Delta h \tag{6}$$

where the term 1/1000 is used to convert units from $l\,s^{-1}$ to $m^3\,s^{-1}$, Ql_h is the total leakage flow at time h and Δh (s) is the time in which Ql_h ($l\,s^{-1}$) is applied.

However, in most of the cases, the leakage flow of the entire network is the only available data. For these analyses, methodologies to estimate the leakage flow at each node from the total leakage rate have been proposed. In this work, the methodology proposed by Araujo et al. [36] was used. From the current leakage rate, which is usually estimated as a percentage of the minimum total night flow, an iterative process was carried out to determine the value of c_i which accomplished Equations (3) and (4) using the software EPANET [35]. This iterative process was required to set the value of c_i at each node and thus determine the leakage flow according to the pressure received at each node [37]. After that, two types of demand were applied to the nodes, one associated with the human consumption, while the other was pressure-dependent and related to leakage. Therefore, a genetic algorithm was applied to determine the new demand pattern which, multiplied by the base demand and added up the leakage flow, matched with the actual measured hourly demand of the network. A detailed description of the above methodology can be found in [36]. Once the leakage flow after installing a PRV or a PAT was obtained, the water cost saving was determined by Equation (5).

The procedure to determine the water saving was carried out in MATLAB [38].

As for the energy cost saving, Equation (7) was used to estimate its value:

$$ECS = \frac{1}{1000} \cdot ndays \cdot \sum_{h=1}^{24} \rho \cdot g \cdot Q_{PAT,h} \cdot H_{PAT,h} \cdot \eta_h \cdot ep \tag{7}$$

where the term 1/1000 is used to convert units from W to kW, ρ is the water density ($kg\,m^{-3}$), g the gravity acceleration ($m\,s^{-2}$), $Q_{PAT,h}$ ($m^3\,s^{-1}$) and $H_{PAT,h}$ (m) the flow and head provided by the PAT at time h, respectively. η_h was the efficiency of the PAT and ep the savings from displaced electricity costs ($0.17 \, € \, kWh^{-1}$ has been assumed in this work, [39]).

Different methodologies to estimate the friction losses in a PAT have been developed [40,41]. A simplified approach similar to the one proposed by [23] was considered in this work. According to [42], the maximum PAT efficiency, i.e., the efficiency of a PAT at the Best Efficiency Point (BEP) is achieved when the flow at the PAT is 75% of the maximum flow, considering 0% as the minimum flow rate through a PAT and 100% as the maximum flow at the same speed. Therefore, two considerations were assumed in this work: (1) $Q_{PAT,h}$ was Q_{BEP} when the input flow was higher than Q_{BEP} and η_h matched with the maximum PAT flow-to-wire efficiency, η_{max}, assumed here as 0.65 [18,25]. In this case, the rest of the flow was bypassed; (2) $Q_{PAT,h}$ matched with the input flow when it was lower than Q_{BEP} and the η_h was determined according to the methodology proposed by [42], in which they evaluated the performance of 113 PATs by characterising the relationship between head and specific speed. The 113 PATs performance database was included in EPANET, to estimate the hydropower potential according to the instant flow and head at a certain site.

A flow chart with a summary of the methodology above described is shown in Figure 1.

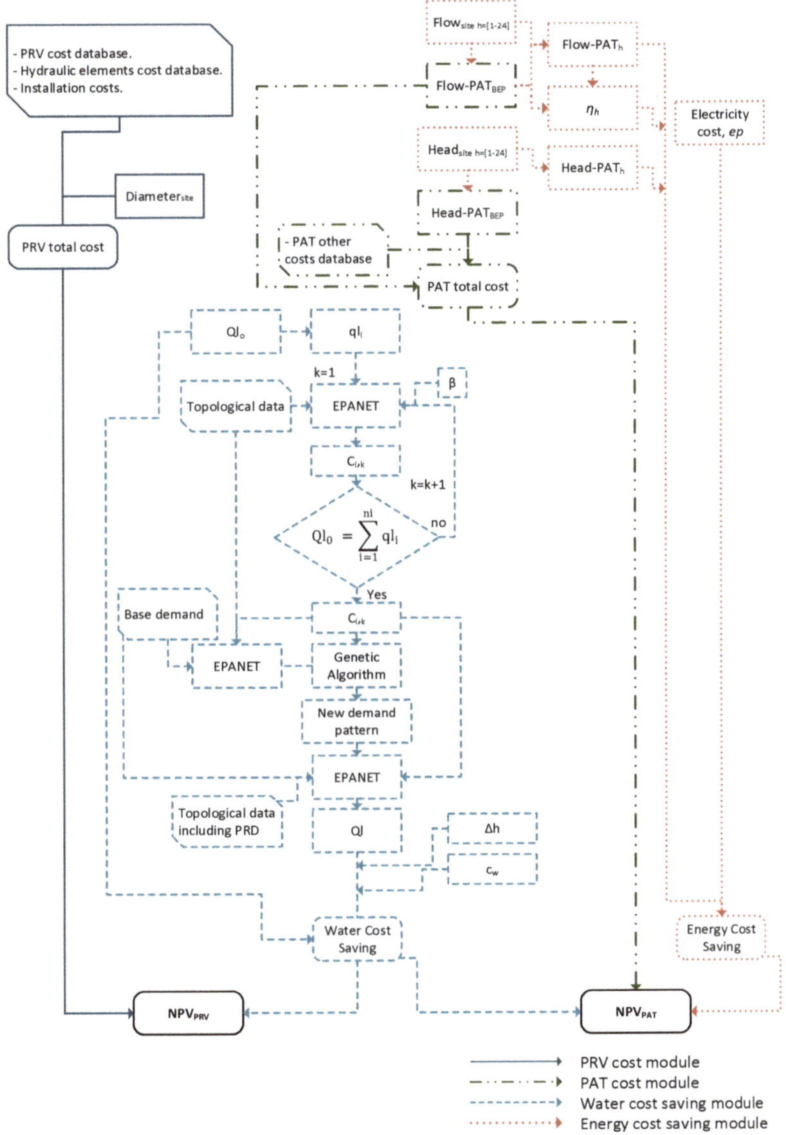

Figure 1. Flow chart of the model to determine the most cost-effective device.

2.2. PRV Total Installation Costs

The installation of a PRV in addition to the valve itself usually requires an upstream isolation valve, a strainer, a downstream isolation valve and an air valve (Figure 2a). In order to select the PRV to be installed in a certain pipe, the flow through it, its diameter and the pressure setting range need to be considered.

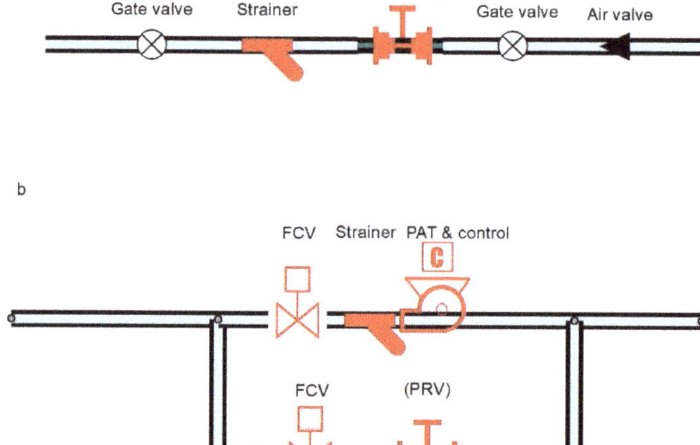

Figure 2. Installation scheme of a Pressure Reducing Valve (PRV) (**a**) and of a Pump As Turbine (PAT) (**b**).

Data related to the cost of PRVs from several manufacturers for a wide range of diameters (50, 65, 80, 100, 125, 150, 200, 250, 300, 400, 500 and 600 mm) and for different pressure settings, were collected [43–51]. Likewise, the costs of the hydraulic elements required in the installation of a PRV for different diameters were gathered. Civil works were also considered, and assumed here as the 10% of the total costs [52].

2.3. PAT Total Installation Cost

The cost in the European context of a centrifugal PAT with a connected four-pole asynchronous motor used as a generator can be estimated as function of its nominal flow and head working conditions at the Best Efficiency Point (BEP) through a set of linear correlations [53]:

$$C_{PAT+gen}(\text{\euro}) = 12,717.29 \cdot Q_{PATBEP} \sqrt{H_{PATBEP}} + 1038.44 \tag{8}$$

The above equation was determined from a database of 343 commercially available pumps and 286 generators [53]. Only asynchronous induction motors which can efficiently work as generators and are commonly sold as standard prime movers of hydraulic pumps were selected. More specifically, four-pole asynchronous motors have been considered as their moderate nominal speed of about 1500 RPM prevents PATs from reaching an excessive speed under runaway conditions, compared to two-poles units.

Apart from the purchase of the turbine and the generator, the final cost of a typical Micro-Hydropower (MHP) scheme includes additional contributions (see Figure 2b) which can be grouped as follows:

- Civil works and hydraulic equipment, including:

    ○   a bypass pipe;
    ○   a set of actuated or manually operated control and sectioning valves;
    ○   a PRV (in certain installations). The necessity of installing this device depends on whether the PAT will operate far from its BEP at a given site, and the capability of the actuated valves to control the pressure received at downstream nodes. If the pressure is too high to be controlled by the actuated control valves, a PRV installed in the bypass is required.
    ○   a Y-strainer;
    ○   a powerhouse hosting the equipment.

- Electric cabinet and control system
- Grid connection fee
- Commissioning
- Other project costs (including consultancy)

The cost of the above elements in PAT energy recovery installations is unclear, since no comprehensive study has provided a cost breakdown of PAT-based MHP schemes embedded in water infrastructures to date. Some authors have suggested that the contribution given by the purchase of a conventional turbine and generator may account typically for the 35% [54] up to a maximum of 70% of the final cost figure of a MHP scheme [18]. However, such figures are not directly applicable to the context of this research since their authors did not consider the use of PATs, which allows for a considerable reduction of the turbine purchase price (5 to 20 times less expensive). In addition, they did not refer specifically to hydropower stations within water distribution networks.

In order to evaluate more accurately the cost of such plants, data from 9 energy recovery schemes in water networks from different countries have been compiled [24,55–57]. All of the selected plants adopted a PAT as a generating device, and had nominal powers ranging from 9 to 120 kW. The location and power rating of all 9 schemes is displayed in Figure 3. According to the available information, it was possible to sub-divide the total cost of the installation into single components:

- turbogenerator alone;
- turbogenerator and control system;
- commissioning;
- civil works and hydraulic equipment;
- grid connection;
- other project costs.

From the purchase price of a PAT and a generator by means of Equation (8), the resulting cost breakdown allowed the realistic quantification of the total expected cost of a PAT-based MHP plant in water infrastructures from its nominal values of flow rate and hydraulic head.

2.4. Case Studies

The first case study consisted of a database with 156 PRVs placed in water infrastructure in Wales and the West Midland regions of the UK (see Figure 4). Information related to average flow, pressure and pipe diameter were collected at those sites [18]. The average flow ranged from $0.2\,l\,s^{-1}$ to $296.5\,l\,s^{-1}$ with associated head values of 32 m and 15 m, respectively. As for diameters, the values were 150 mm, 200 mm, 250 mm, 300 mm and 400 mm.

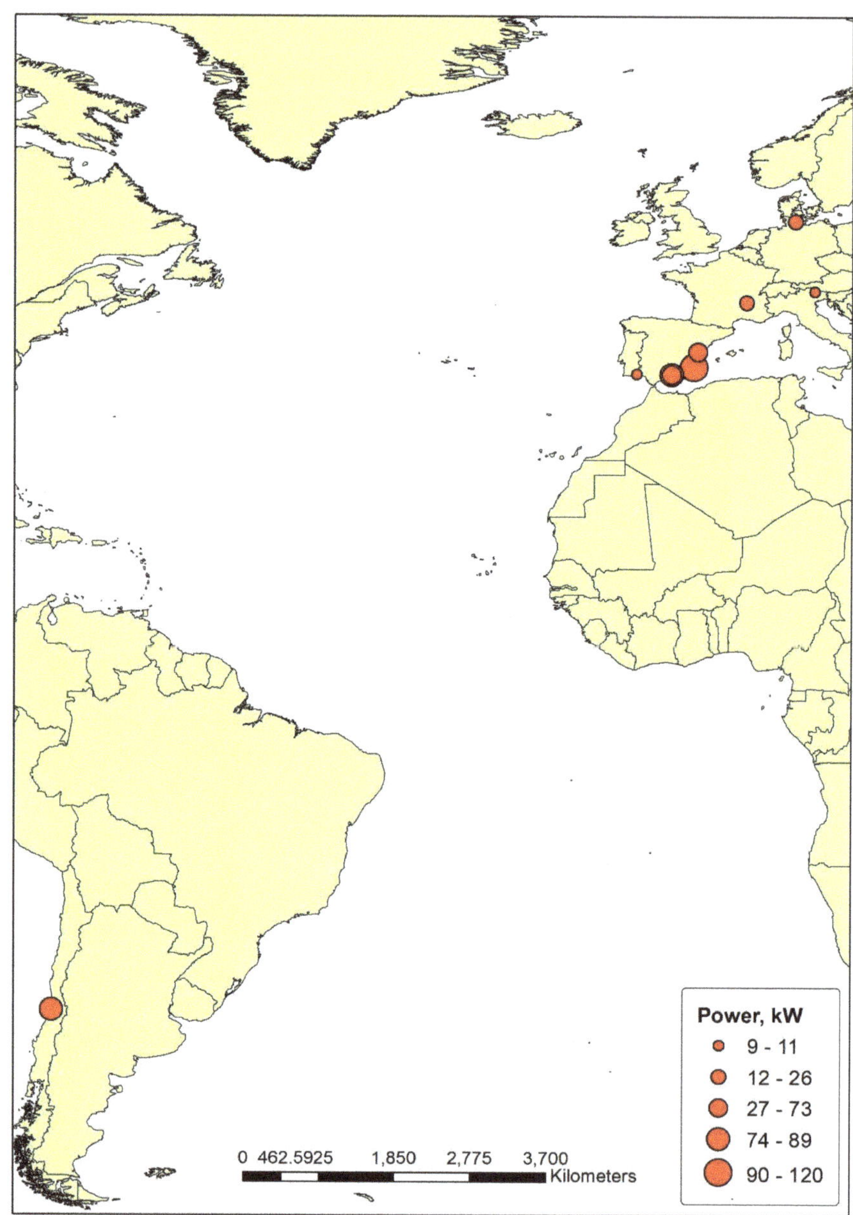

Figure 3. Location of the PAT plants used to estimate the total PAT cost.

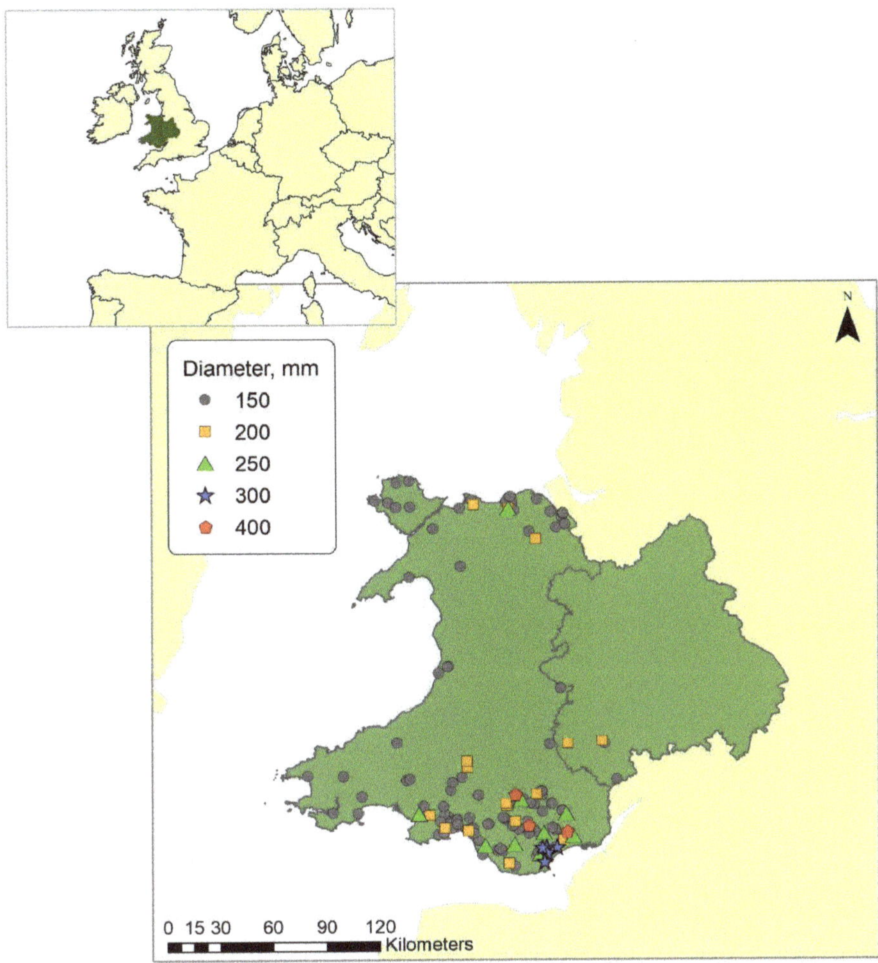

Figure 4. Location of the selected PRV sites.

The second case study focused on Ballinabranna Group Water Scheme (BGWS), a water supply network located in a rural area in Ireland. BGWS is a gravity network which supplies an average demand of 6.16 l s^{-1} to 44 consumption nodes. This network covers an area of almost 54 km^2, with diameters ranging from 50 mm to 150 mm. Three PRVs are currently operating in the network. However, in an analysis to determine the optimum PAT locations in this network, a new potential site was detected [37]. This site showed higher MHP potential than that of the current PRV locations. Hence, the analysis carried out in this work took into account the determination of the NPV in two of the current PRVs and at the proposed additional site (site 3 in Figure 5).

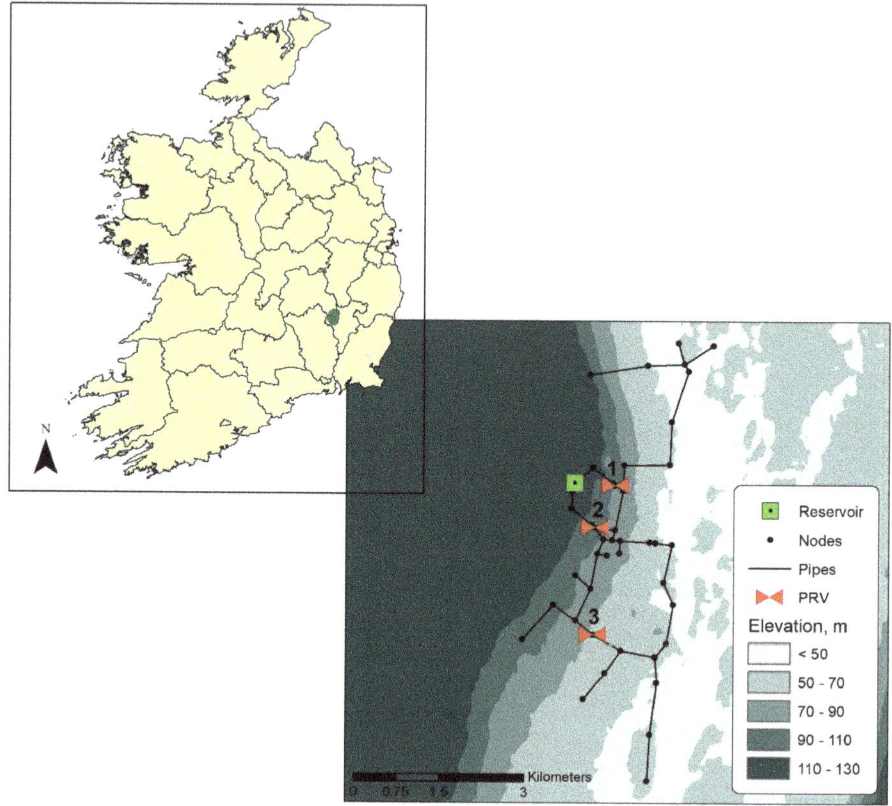

Figure 5. Ballinabranna water supply network layout.

3. Results

3.1. PRV Cost Model

The cost of 110 PRVs from different manufacturers were collected. The diameter of the selected PRVs ranged from 50 mm to 600 mm while the pressure setting range was 6–60 m, 15–60 m, 10–100 m 10–160 m, 14–72 m, 40–120 m, 50–120 m, 10–70 m, 17–86 m and 7–70 m. The average PRV cost was 3516 € with a minimum value of 162 € for a 50 mm diameter and a maximum price of 31,765 € for a 500 mm diameter (see Figure 6). As could be expected, the greater the diameter the higher the PRV cost. However, a clear difference between the range of the pressure setting and the PRV cost was not detected. The differences in the PRV cost were mainly related to the different manufacturers who provided the data.

To estimate the total cost, the cost of the two required gate valves, the strainer and the air valve were also considered (Table 1). The cost of the two gate valves ranged from 146 € to 6370 € for diameters of 50 mm and 600 mm respectively, whereas the strainer cost varied from 148 € to 18,063 € for the same diameters. As for the air valve, the cost ranged from 543 € for the 50 mm diameter to 19,286 € for the 600 mm diameter.

These additional costs and the amount of civil works were added to the PRV cost to estimate the total installation cost. Considering an average PRV cost for each diameter according to the values provided by the different manufacturers, the average total cost was 18,762 €, ranging from 1738 € for a 50 mm diameter pipeline to 72,621 € for a 600 mm diameter (see Figure 7). When the minimum PRV cost for each diameter was taken into account the average total cost was 16,931 €, varying from 1099 € to 72,621 € for diameters of 50 mm and 600 mm. As for the total cost when the maximum PRV cost for each diameter was selected, the average value was 21,258 € ranging from 2520 € to 72,621 € for diameters of 50 mm and 600 mm.

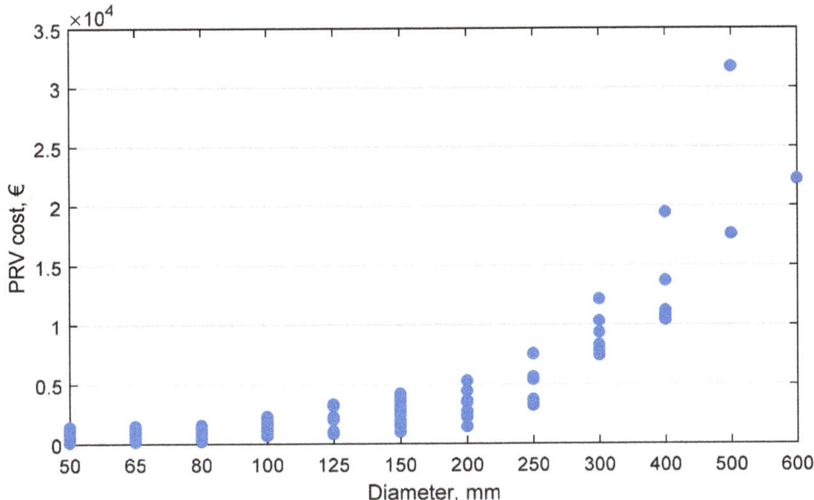

Figure 6. PRV cost for diameters ranging from 50 mm to 600 mm.

Table 1. Additional costs related to the PRV installation.

	Diameter, mm											
	50	**65**	**80**	**100**	**125**	**150**	**200**	**250**	**300**	**400**	**500**	**600**
Gate valve (2)	146	190	218	274	374	514	800	1216	1676	2899	4462	6370
Strainer	148	159	208	227	410	391	717	1377	2104	5331	11,669	18,063
Air valve	543	543	615	854	1215	1290	2520	4430	4947	8953	13,624	19,286
TOTAL	837	892	1041	1355	1999	2195	4037	7023	8727	17,183	29,756	43,719

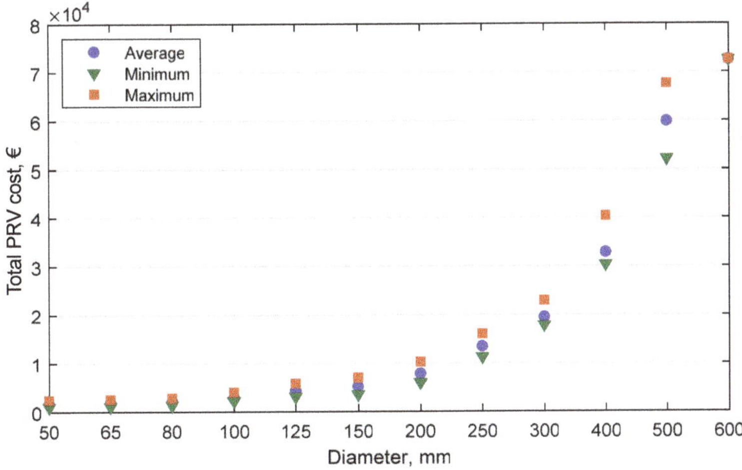

Figure 7. Total PRV cost for diameters ranging from 50 mm to 600 mm.

3.2. PAT Cost Model

The model developed by [53] was used to determine the PAT cost. A flow range between $0.01{\cdot}10^{-3}$ m^3 s^{-1} and 0.4 m^3 s^{-1} and pressure values varying from 5 m to 100 m were considered. The average PAT cost was 18,666 €, ranging from 1039 € for the minimum values of flow and pressure to 51,908 € associated with a flow of 0.4 m^3 s^{-1} and a pressure of 100 m (see Figure 8).

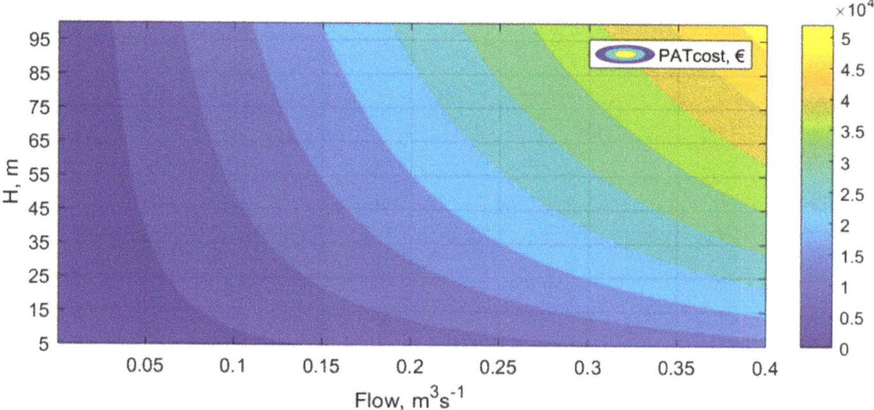

Figure 8. PAT cost according to flow and head.

The analysis of the cost components from the 9 evaluated PAT schemes according to the categories listed in Section 2.3, did not lead to any identifiable trends according to the nominal power of the scheme. Instead, it is reasonable to assume that a number of site-specific conditions (e.g., amount of civil works required, distance from the electric grid, and effectiveness of the design) resulted in varied percentages of each analysed cost category. The overall results of the analysis are shown in Figure 9, where the vertical error bars indicate the minimum and maximum percentages for each cost category.

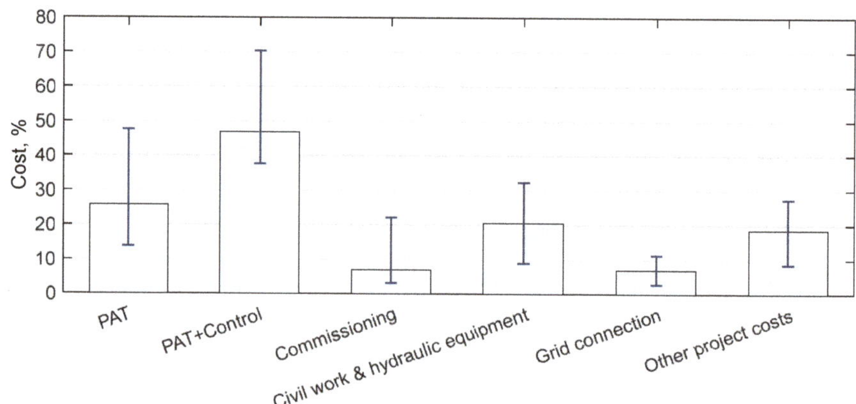

Figure 9. Costs involved in the PAT installation.

The results from Figure 9 showed that across the 9 analysed schemes the purchase of the PAT and generator accounted for 26% of the total cost of the installation on average, ranging from 14% to 48% (see Figure 9). Hence, the results from Equation (8) were divided by a factor of 0.26 in order to achieve a realistic estimation of the total expected cost for a PAT-based energy recovery scheme according to its mean operating conditions (Figure 10).

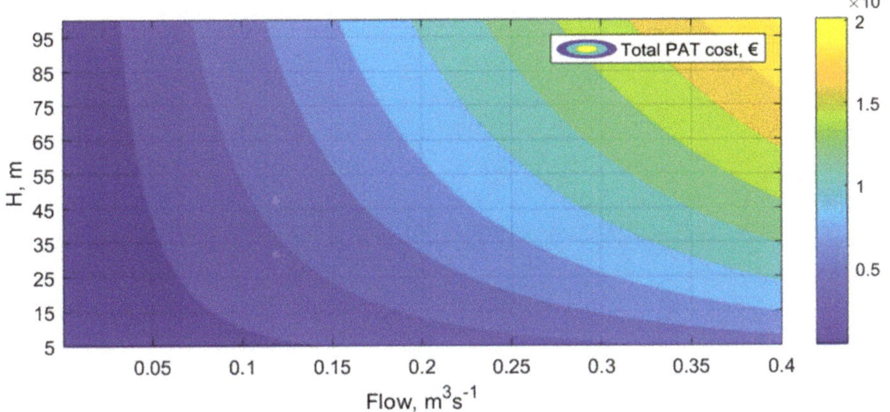

Figure 10. Total PAT cost according to flow and head.

Apart from the turbine and generator, the PAT control devices (with an average cost share of 24%) the civil work and hydraulic equipment (involving an average of 21% of the total cost) and the other project costs (accounting for 19%) were found to be the most expensive elements in a typical PAT installation.

Considering the PAT cost as 26% of the total cost, the total PAT installation cost was estimated for the flow range $0.01 \cdot 10^{-3}$ m^3 s^{-1}–0.4 m^3 s^{-1} and pressure values from 5 m to 100 m. The average total PAT cost was 72,518 €, varying from 4035 € to 201,662 €.

When the PAT cost was assumed as 48% of the total cost, the average PAT installation cost was 39,214 €, with a minimum cost of 2182 € for the lowest values of flow and pressure, and a maximum cost of 109,050 €. Taking the PAT cost as the 14% of the total cost, instead, the average total cost was 134,288 €, varying from 7473 € to 373,436 €.

3.3. Determination of the Most Feasible Pressure Reduction Device, PRV or PAT, in the Case Studies

For case study 1, only average flow and head data were available. Hence, the generated power was determined considering that the average values matched with the flow and head of the PAT at the BEP. 140 sites out of 156 showed an average power higher than the minimum threshold value (400 W), with an average value of 4338 W, ranging from 460 W to 31,455 W (see Table 2). As for the total installation cost, the PRV cost was related to the pipe diameter while the PAT cost varied according to the generated power (see Figure 11).

Table 2. PRV/ PAT analysis in case study 1.

	Power, kW	Total PRV Cost, €	Total PAT Cost, €			PAT NPV, €			PAT Payback Period, years		
			26% [1]	48% [2]	14% [3]	26% [1]	48% [2]	14% [3]	26% [1]	48% [2]	14% [3]
Ave.	4338	7236	10,397	5567	20,831	57,960	61,489	59,163	4	3	6
Max.	31,455	32,909	60,770	32,862	112,534	451,467	467,421	421,875	10	7	10
Min.	460	5284	4761	2575	9097	2695	4197	5079	2	2	3

[1] Considering the PAT cost as 26% of the total cost. [2] PAT cost as 48% of the total cost. [3] PAT cost as 14% of the total cost.

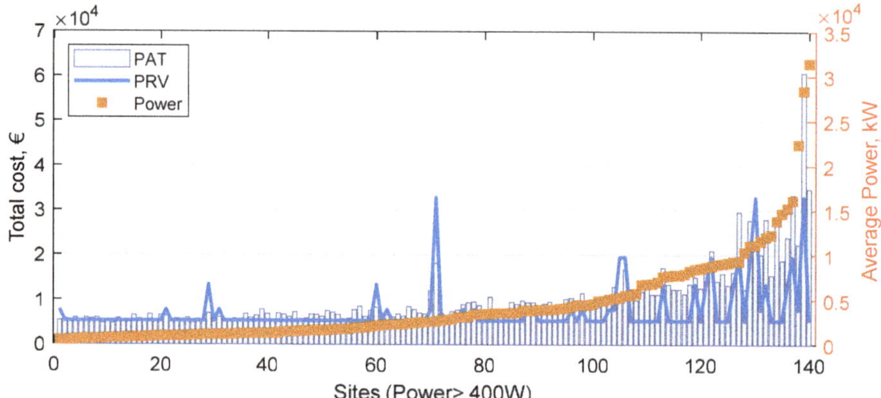

Figure 11. PAT vs. PRV total cost according to the average power at the evaluated sites.

Considering in this case the average total cost of both PRV and PAT, the PRV was the less expensive option in the majority of the sites, with an average total cost of 7236 € (see Table 2). In only 15 sites (10%) the PAT showed lower capital costs. Analysing these sites, the PAT total cost was slightly cheaper in 6 sites with a 150 mm diameter and power lower than 1 kW. When the generated power was higher the PRV was the best option considering this diameter. As for the 200 mm diameters, the PAT was the best option in 3 sites with an average power lower than 2300 W. When the generated power at the sites with a 200 mm diameter was higher, the PRV was cheaper. A similar tendency was found for pipes of 250, 300 and 400 mm. The lower the generated power the higher the probability that the PAT would be less expensive (see Figure 11).

As for the economic analysis in case study 1, only the energy cost saving was estimated since information about the effects of the PRV performance on nodes pressure was not available. The NPV was determined considering the energy that could be generated in each evaluated site during the PAT lifespan (15 years). Considering that the PAT cost was 26% of the total cost, the NPV showed positive values in all the 140 sites. The estimated average NPV was 57,960 €, ranging from 2695 € to 451,467 € (see Figure 12). The average payback period was 4 years (see Table 2). Under this assumption, three sites showed payback periods higher than the threshold established (see Figure 13).

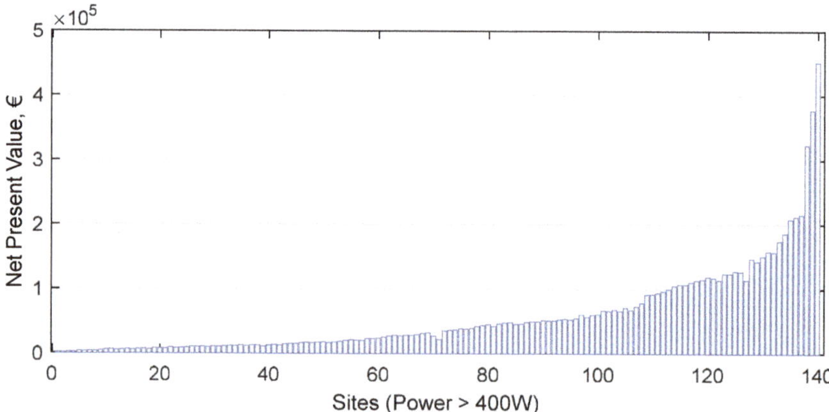

Figure 12. Net present value in the evaluated sites considering that the PAT represents 26% of the total installation cost.

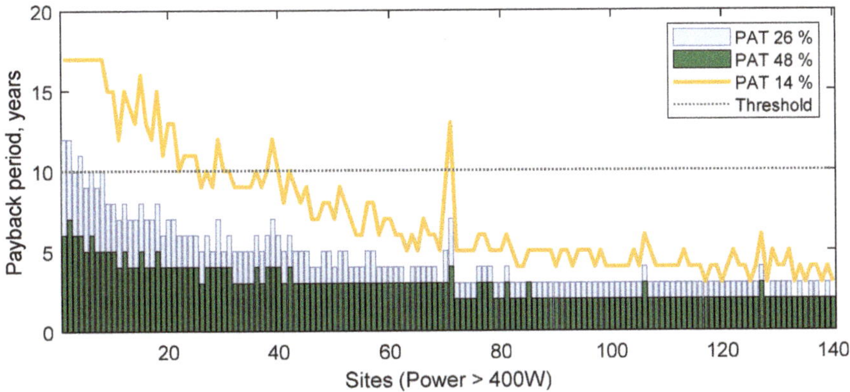

Figure 13. Payback period in the evaluated sites.

When the PAT cost was considered as 14% of the total cost, 113 out of the 140 sites with power higher than 400 W showed a payback period lower than 10 years, with an average payback period of 6 years (see Figure 13). As for the NPV, under this assumption the average value was 59,163 €, varying from 5079 € to 421,875 € (see Table 2). Finally, when the PAT cost was assumed as 48% of the total cost, the payback period was lower than 10 years in all 140 sites with average power higher than 400 W (see Figure 13), showing an average payback period of 3 year. In this case, the average NPV was 61,489 € with a minimum and a maximum value of 4197 € and 467,421 € (see Table 2). Considering the evaluated sites in case study 1, the PAT installation costs could be even lower since a current PRV is already operating, making the PAT installation as a retro-fit, more feasible.

For case study 2, the replacement of the three PRVs with PATs would involve powers at the BEP of 1223 W, 1533 W and 457 W (see Table 3). As for the total installation cost, the PRV showed the lowest values in the majority of the cases. The PAT installation cost was only lower at site 1, with a diameter of 150 mm, when the PAT cost was assumed as the 48% of the PAT total cost. The average installation cost ranged from 3009 € to 5284 € when the PRV was considered and from 5323 € to 6268 € for the PAT installation.

Table 3. PRV/ PAT analysis in Ballinabranna Group Water Scheme (BGWS).

	PRV			PAT		
Site Number	**1**	**2**	**3**	**1**	**2**	**3**
Diameter, mm	150	100	100	150	100	100
Power BEP, W				1223	1533	457
Energy, kWh year^{-1}				7346	9203	2744
Ave cost, €	5284	3009	3009	5919	6268	5323
Min cost, €	3574	2303	2303	3201	3390	2879
Max cost, €	7074	4067	4067	10,984	11,633	9879
Water saving, m^3 year^{-1}	15,559	15,559	15,559	15,559	15,559	15,559
Water cost saving, € year^{-1}	4668	4668	4668	4668	4668	4668
Energy cost saving, € year^{-1}				1249	1565	466
Payback period, year	3.0	2.0	2.0	3.0	3.0	3.0
NPV, €	43,164	45,440	45,440	55,493	58,420	47,967
NPV/Water saving, € m^{-3}	0.18	0.19	0.19	0.24	0.25	0.21

The economic analysis in case study 2 was carried out considering both the energy cost saving and the water saving as a result of the PRV/PAT operation. As initial leakage rate, 80% of the minimum night-time flow was assumed [36,58]. The same pressure drop was considered for the PRVs and the

PATs. Hence, the water saving and the water cost saving showed the same values for both devices: 15,559 m^3 year^{-1} and 4668 € year^{-1} per pressure reducing device (see Table 3). The operation of the three PATs would involve an annual energy cost saving of 1249 €, 1565 € and 466 €. Thus, the NPV considering the PAT lifespan and the average installation cost showed values of 55,493 €, 58,420 € and 47,967 € at sites 1, 2 and 3, respectively. This was 29%, 29% and 6% higher than the estimated NPV considering the average PRV installation cost. The ratio between the NPV and the water leakage that could be saved during the PAT/ PRV lifespan offered values for the PATs of 0.24 € m^{-3}, 0.25 € m^{-3} and 0.21 € m^{-3}, again higher than the estimated values for the PRVs (see Table 3).

Therefore, this work shows that although the required investment for a PRV installation is lower in most cases, the energy and hence, the cost savings that could be achieved by considering a PAT justify its installation. While both devices are equally efficient in terms of leakage reduction, the additional energy cost savings achieved by the PAT performance make its installation as a device to reduce leakage and generate energy more feasible.

4. Discussion

PATs are considered to be a cheaper technology compared to traditional turbines for small hydropower energy recovery [26,59]. However, information related to the total PAT cost, also including installation cost, is not easily accessible. Overall, methodologies focused on the use of PATs took into account a PAT cost according to the generated power. Thus, [60] considered an average PAT investment cost of 545 € kW^{-1} for generated power lower than 10 kW, including the PAT cost and the civil, electrical and electronic equipment. [25] determined the total installation cost considering an average PAT cost related to the generated power of 220 € kW^{-1} plus two additional costs of 450 € and 2500 € (associated to piping and grid connection). Other methodologies established the installation cost either as a percentage of the total installation cost or as a cost associated to the generated power. [30] considered the installation cost as 65% of the total cost, whereas [23] proposed an installation cost of 350 € kW^{-1}.

The methodology presented in this work analysed the installation cost and categories of 9 MHP plants in different countries, thus determining an average installation cost and the range. The installation cost determined in the sites of case study 1, using the methodology presented in this work compared with previous methodologies, showed higher total PAT costs using the values obtained from the 9 MHP plants. Thus, the average total PAT cost at the sites in case study 1 was 3904 €, 2364 €, 4159 € and 4251 € applying the methodologies proposed by [23,25,30,60], respectively. These were all considerably lower than the average total value presented in this work, 10,397 €.

Above all, the model presented here shows that when the total costs and benefits of the installation of PATs or PRVs as pressure reduction devices are compared, the PAT is the more feasible option with greater long term benefits. These greater long term benefits were also consistently found when a sensitivity analysis of the assumed variables was conducted (i.e., using min, average and maximum values for percentage contribution of differing cost components).

5. Conclusions

A model to determine the most feasible device, PRV or PAT, considering all the costs involved in a PRV and a PAT installation, as well as the water cost savings associated to leakage reduction, and the energy cost saving related to the PAT operation has been proposed. This work highlights the potential of MHP installations in general and PATs in particular as a tool to improve the sustainability of water distribution networks and reduce electricity costs. The analysis carried out showed that a wider database with the real PAT installation cost would be required to better estimate the feasibility of the installation of these devices. Limitations of the current methodology include: The limited amount of information available on PAT costs at very low powers; the lack of significant amounts of case studies on which to base estimates of civil works and other project costs; and the lack of information on the cost and performance of PAT types other than centrifugal.

The main finding of the work is that the use of a PRV over a PAT as a pressure reduction tool is not the most economically viable or sustainable option. While installing a PRV reduces leakage and thus energy consumption, it wastes the opportunity to recover electricity in the network. The development of less expensive turbines and/or pump as turbines adapted to work under low flow-head condition will tip the balance toward the installation of these devices even more.

Author Contributions: A.M.N., I.F.G. and D.N. conceived and developed the model; I.F.G. and D.N. performed the simulations and analyzed the results; I.F.G. and D.N. wrote the paper; A.M.N., I.F.G. and D.N. revised and corrected the final document.

Funding: This study was part funded by the ERDF Interreg Ireland-Wales programme 2014–2020 through the Dwr Uisce project.

Conflicts of Interest: The authors declare no conflict of interest.

References

1. WWAP (United Nations World Water Assessment Programme). *The United Nations World Water Development Report 2018: Nature-Based Solutions for Water*; WWAP: Paris, France, 2018.
2. Singh, L.K.; Jha, M.K.; Chowdary, V.M. Multi-criteria analysis and GIS modeling for identifying prospective water harvesting and artificial recharge sites for sustainable water supply. *J. Clean. Prod.* **2017**, *142*, 1436–1456. [CrossRef]
3. McDonald, R.I.; Weber, K.; Padowski, J.; Flörke, M.; Schneider, C.; Green, P.A.; Gleeson, T.; Eckman, S.; Lehner, B.; Balk, D.; et al. Water on an urban planet: Urbanization and the reach of urban water infrastructure. *Glob. Environ. Chang.* **2014**, *27*, 96–105. [CrossRef]
4. European Environment Agency. Use of Freshwater Resources. Available online: https://www.eea.europa.eu/data-and-maps/indicators/use-of-freshwater-resources-2/assessment-2 (accessed on 24 May 2018).
5. U.S. Energy Information Administration. International Energy Outlook. 2017. Available online: www.eia.gov/ieo (accessed on 24 May 2018).
6. Enerdata. Enerdata World Energy Consumption Statistics. Available online: https://yearbook.enerdata.net/total-energy/world-consumption-statistics.html (accessed on 14 February 2018).
7. World-Bank. What is Non-Revenue Water? How Can We Reduce It for Better Water Service? Available online: https://blogs.worldbank.org/water/what-non-revenue-water-how-can-we-reduce-it-better-water-service (accessed on 5 June 2019).
8. Loureiro, D.; Amado, C.; Martins, A.; Vitorino, D.; Mamade, A.; Coelho, S.T. Water distribution systems flow monitoring and anomalous event detection: A practical approach. *Urban. Water J.* **2016**, *13*, 242–252. [CrossRef]
9. Mérida García, A.; Fernández García, I.; Camacho Poyato, E.; Montesinos Barrios, P.; Rodríguez Díaz, J.A. Coupling irrigation scheduling with solar energy production in a smart irrigation management system. *J. Clean. Prod.* **2018**, *175*, 670–682. [CrossRef]
10. Kollmann, R.; Neugebauer, G.; Kretschmer, F.; Truger, B.; Kindermann, H.; Stoeglehner, G.; Ertl, T.; Narodoslawsky, M. Renewable energy from wastewater—Practical aspects of integrating a wastewater treatment plant into local energy supply concepts. *J. Clean. Prod.* **2017**, *155*, 119–129. [CrossRef]
11. Gupta, A.; Dpt, C.; National, V.; Bokde, N.; Dpt, C.; National, V.; Marathe, D.; Dpt, C.; National, V.; Kulat, K.; et al. Leakage Reduction in Water Distribution Systems with Efficient Placement and Control of Pressure Reducing Valves Using Soft Computing Techniques. *Eng. Technol. Appl. Sci. Res.* **2017**, *7*, 1528–1534.
12. Parra, S.; Krause, S.; Krönlein, F.; Günthert, F.W.; Klunke, T. Intelligent pressure management by pumps as turbines in water distribution systems: Results of experimentation. *Water Sci. Technol. Water Supply* **2017**, *6*, ws2017154. [CrossRef]
13. Fontana, N.; Giugni, M.; Glielmo, L.; Marini, G.; Zollo, R. Real-Time Control of Pressure for Leakage Reduction in Water Distribution Network: Field Experiments. *J. Water Resour. Plan. Manag.* **2018**, *144*, 04017096. [CrossRef]
14. Gallagher, J.; Styles, D.; McNabola, A.; Williams, A.P. Life cycle environmental balance and greenhouse gas mitigation potential of micro-hydropower energy recovery in the water industry. *J. Clean. Prod.* **2015**, *99*, 152–159. [CrossRef]

15. Lydon, T.; Coughlan, P.; McNabola, A. Pressure management and energy recovery in water distribution networks: Development of design and selection methodologies using three pump-as-turbine case studies. *Renew. Energy* **2017**, *114*, 1038–1050. [CrossRef]

16. McNabola, A.; Coughlan, P.; Corcoran, L.; Power, C.; Prysor Williams, A.; Harris, I.; Gallagher, J.; Styles, D. Energy recovery in the water industry using micro-hydropower: An opportunity to improve sustainability. *Water Policy* **2014**, *16*, 168. [CrossRef]

17. Power, C.; Coughlan, P.; McNabola, A. Microhydropower Energy Recovery at Wastewater-Treatment Plants: Turbine Selection and Optimization. *J. Energy Eng.* **2017**, *143*, 04016036. [CrossRef]

18. Gallagher, J.; Harris, I.M.; Packwood, A.J.; McNabola, A.; Williams, A.P. A strategic assessment of micro-hydropower in the UK and Irish water industry: Identifying technical and economic constraints. *Renew. Energy* **2015**, *81*, 808–815. [CrossRef]

19. Sinagra, M.; Sammartano, V.; Morreale, G.; Tucciarelli, T. A New Device for Pressure Control and Energy Recovery in Water Distribution Networks. *Water* **2017**, *9*, 309. [CrossRef]

20. Samora, I.; Manso, P.; Franca, M.; Schleiss, A.; Ramos, H. Energy Recovery Using Micro-Hydropower Technology in Water Supply Systems: The Case Study of the City of Fribourg. *Water* **2016**, *8*, 344. [CrossRef]

21. Sammartano, V.; Sinagra, M.; Filianoti, P.; Tucciarelli, T. A Banki–Michell turbine for in-line water supply systems. *J. Hydraul. Res.* **2017**, *55*, 686–694. [CrossRef]

22. Sinagra, M.; Aricò, C.; Tucciarelli, T.; Amato, P.; Fiorino, M.; Sinagra, M.; Aricò, C.; Tucciarelli, T.; Amato, P.; Fiorino, M. Coupled Electric and Hydraulic Control of a PRS Turbine in a Real Transport Water Network. *Water* **2019**, *11*, 1194. [CrossRef]

23. Carravetta, A.; del Giudice, G.; Fecarotta, O.; Ramos, H. PAT Design Strategy for Energy Recovery in Water Distribution Networks by Electrical Regulation. *Energies* **2013**, *6*, 411–424. [CrossRef]

24. Carravetta, A.; Derakhshan Houreh, S.; Ramos, H.M. *Application of PAT Technology*; Springer International Publishing: Berlin/Heidelberg, Germany, 2018; pp. 189–218.

25. Fecarotta, O.; McNabola, A. Optimal Location of Pump as Turbines (PATs) in Water Distribution Networks to Recover Energy and Reduce Leakage. *Water Resour. Manag.* **2017**, *31*, 5043–5059. [CrossRef]

26. Novara, D.; Carravetta, A.; Derakhshan, S.; McNabola, A.; Ramos, H.M. A Cost model for Pumps as Turbines and a comparison of design strategies for their use as energy recovery devices in Water Supply Systems. In Proceedings of the 10th International Conference on Energy Efficiency in Motor Driven Systems (EEMODS'17), Rome Italy, 6–8 September 2017.

27. Giugni, M.; Fontana, N.; Ranucci, A. Optimal Location of PRVs and Turbines in Water Distribution Systems. *J. Water Resour. Plan. Manag.* **2014**, *140*, 1–6. [CrossRef]

28. Tricarico, C.; Morley, M.S.; Gargano, R.; Kapelan, Z.; Savić, D.; Santopietro, S.; Granata, F.; de Marinis, G. Optimal energy recovery by means of pumps as turbines (PATs) for improved WDS management. *Water Sci. Technol. Water Supply* **2018**, *184*, 1365–1374. [CrossRef]

29. Pérez-Sánchez, M.; Sánchez-Romero, F.; Ramos, H.; López-Jiménez, P.A. Optimization Strategy for Improving the Energy Efficiency of Irrigation Systems by Micro Hydropower: Practical Application. *Water* **2017**, *9*, 799. [CrossRef]

30. García Morillo, J.; McNabola, A.; Camacho, E.; Montesinos, P.; Rodríguez Díaz, J.A. Hydro-power energy recovery in pressurized irrigation networks: A case study of an Irrigation District in the South of Spain. *Agric. Water Manag.* **2018**, *204*, 17–27. [CrossRef]

31. Fecarotta, O.; Aricò, C.; Carravetta, A.; Martino, R.; Ramos, H.M. Hydropower Potential in Water Distribution Networks: Pressure Control by PATs. *Water Resour. Manag.* **2015**, *29*, 699–714. [CrossRef]

32. Patelis, M.; Kanakoudis, V.; Gonelas, K. Pressure Management and Energy Recovery Capabilities Using PATs. *Procedia Eng.* **2016**, *162*, 503–510. [CrossRef]

33. The Public Spending Code. Ireland Department of Public Expenditure and Reform E-Technical References. Available online: http://publicspendingcode.per.gov.ie/technical-references/ (accessed on 14 May 2018).

34. Araujo, L.S.; Ramos, H.; Coelho, S.T. Pressure Control for Leakage Minimisation in Water Distribution Systems Management. *Water Resour. Manag.* **2006**, *20*, 133–149. [CrossRef]

35. Rossman, L.A. *EPANET 2 Users Manual EPA/600/R-00/57*; U.S. Environmental Protection Agency: Cincinnati, OH, USA, 2000.

36. Araujo, L.S.; Ramos, H.; Coelho, S.T. Estimation of distributed pressure-dependent leakage and consumer demand inwater supply networks. In *Advance in Water Supply Management*; CRC Press: London, UK, 2003.

37. Fernández García, I.; Ferras, D.; McNabola, A. Potential of energy recovery and water saving using microhydropower in rural water distribution networks. *J. Water Resour. Plan. Manag.* **2019**, *145*, 1–11. [CrossRef]

38. The MathWorks Inc. *MATLAB Version R2016b*; The MathWorks Inc.: Natick, MA, USA, 2016.

39. SEAI. Electricity & Gas Prices in Ireland. 1st Semester (January–June) 2017. Available online: https://www.seai.ie/resources/publications/Electricity_Gas_Prices_January_June_2017 (accessed on 10 May 2018).

40. Carravetta, A.; Fecarotta, O.; Ramos, H.M. A new low-cost installation scheme of PATs for pico-hydropower to recover energy in residential areas. *Renew. Energy* **2018**, *125*, 1003–1014. [CrossRef]

41. Fecarotta, O.; Ramos, H.M.; Derakhshan, S.; Del Giudice, G.; Carravetta, A. Fine Tuning a PAT Hydropower Plant in a Water Supply Network to Improve System Effectiveness. *J. Water Resour. Plan. Manag.* **2018**, *144*, 04018038. [CrossRef]

42. Novara, D.; McNabola, A. A model for the extrapolation of the characteristic curves of Pumps as Turbines from a datum Best Efficiency Point. *Energy Convers. Manag.* **2018**, *174*, 1–7. [CrossRef]

43. Grancho Ferreira, A.R. Energy Recovery in Water Distribution Networks towards Smart Water Grids. Master's Thesis, University of Lisbon, Lisbon, Portugal, 2017.

44. BERMAD PRV cost. Personal Communication, 2018.

45. CLA-VAL PRV cost. Personal Communication, 2018.

46. WATTs PRV cost. Personal Communication, 2018.

47. Flomatic PRV cost. Personal Communication, 2018.

48. AVK PRV cost. Personal Communication, 2018.

49. CSA PRV cost. Personal Communication, 2018.

50. Valves-Online PRV cost. Personal Communication, 2018.

51. Talis PRV cost. Personal Communication, 2018.

52. TRAGSA Tarifas 2015 Para Encomiendas Sujetas a Impuestos. 2015. Available online: https://www.tragsa.es/es/grupo-tragsa/regimen-juridico/Documents/ACTUALIZACI%C3%93N%20TARIFAS%20AGOSTO/Tarifas%202015%20para%20encomiendas%20sujetas%20a%20impuestos.pdf (accessed on 20 June 2019).

53. Novara, D.; Carravetta, A.; McNabola, A.; Ramos, H.M. Cost model for Pumps As Turbines in run-off-river and in-pipe micro-hydropower applications. *J. Water Resour. Plan. Manag.* **2019**, *145*, 04019012. [CrossRef]

54. Ogayar, B.; Vidal, P.G. Cost determination of the electro-mechanical equipment of a small hydro-power plant. *Renew. Energy* **2009**, *34*, 6–13. [CrossRef]

55. Kramer, M.; Terheiden, K.; Wieprecht, S. Pumps as turbines for efficient energy recovery in water supply networks. *Renew. Energy* **2018**, *122*, 17–25. [CrossRef]

56. Livramento, J.M. Central micro-hídrica incorporada em adutora. Master's Thesis, Faculdade de Ciências e Tecnologia, Universidade do Algarve, Faro, Portugal, 2013.

57. Lledó, J. Tecnoturbines. Powering Water 2016. Available online: https://www.mapa.gob.es/images/es/tt-center_tcm30-131554.pdf (accessed on 20 June 2019).

58. Alkasseh, J.M.A.; Adlan, M.N.; Abustan, I.; Aziz, H.A.; Hanif, A.B.M. Applying Minimum Night Flow to Estimate Water Loss Using Statistical Modeling: A Case Study in Kinta Valley, Malaysia. *Water Resour. Manag.* **2013**, *27*, 1439–1455. [CrossRef]

59. Ramos, H.; Borga, A. Pumps as turbines: An unconventional solution to energy production. *Urban. Water* **1999**, *1*, 261–263. [CrossRef]

60. Pérez-Sánchez, M.; Sánchez-Romero, F.J.; López-Jiménez, P.A.; Ramos, H.M. PATs selection towards sustainability in irrigation networks: Simulated annealing as a water management tool. *Renew. Energy* **2018**, *116*, 234–249. [CrossRef]

© 2019 by the authors. Licensee MDPI, Basel, Switzerland. This article is an open access article distributed under the terms and conditions of the Creative Commons Attribution (CC BY) license (http://creativecommons.org/licenses/by/4.0/).

Article

Effect of a Commercial Air Valve on the Rapid Filling of a Single Pipeline: a Numerical and Experimental Analysis

Óscar E. Coronado-Hernández [1,*], Mohsen Besharat [2], Vicente S. Fuertes-Miquel [3] and Helena M. Ramos [2]

1   Facultad de Ingeniería, Universidad Tecnológica de Bolívar, Cartagena 131001, Colombia
2   Department of Civil Engineering and Architecture, CERIS, Instituto Superior Técnico, University of Lisbon, Lisbon 1049–001, Portugal
3   Departamento de Ingeniería Hidráulica y Medio Ambiente, Universitat Politècnica de València, Valencia 46022, Spain
*   Correspondence: ocoronado@utb.edu.co; Tel.: +57-301 3715398

Received: 2 August 2019; Accepted: 29 August 2019; Published: 31 August 2019

Abstract: The filling process in water pipelines produces pressure surges caused by the compression of air pockets. In this sense, air valves should be appropriately designed to expel sufficient air to avoid pipeline failure. Recent studies concerning filling maneuvers have been addressed without considering the behavior of air valves. This work shows a mathematical model developed by the authors which is capable of simulating the main hydraulic and thermodynamic variables during filling operations under the effect of the air valve in a single pipeline, which is based on the mass oscillation equation, the air–water interface, the polytropic equation of the air phase, the air mass equation, and the air valve characterization. The mathematical model is validated in a 7.3-m-long pipeline with a 63-mm nominal diameter. A commercial air valve is positioned in the highest point of the hydraulic installation. Measurements indicate that the mathematical model can be used to simulate this phenomenon by providing good accuracy.

Keywords: air valve; air–water interface; filling; flow; pipelines; transient

---

1. Introduction

Entrapped air inside a liquid in a pressurized pipeline system the cause of numerous serious problems in conveying systems. Specifically, the problem tends to be significantly more substantial when a transient flow scenario exists in the system. The transient flow condition can appear as a consequence of different causes, namely a water hammer event, emptying or filling of a pipeline, pump shut down or start up, the existence of leaks, rapid valve maneuvers, and cavitation occurrence. The aforementioned transient events have been studied extensively in previous studies using different numerical models [1]. The valve maneuver is a crucial issue in pressurized pipelines, in which operating a valve without enough care, might create irretrievable accidents. Azoury et al. [2] studied the effect of a valve closure on water hammer using the method of characteristics. The water hammer numerical simulation in conjunction with a column separation occurrence, was studied by Himr [3], and Simpson and Wylie [4]. Saemi et al. [5] presented a study on two- and three-dimensional calculations of the water hammer flows using computational fluid dynamics (CFD) models. In addition, different studies present the water control issue [6–8]. Among the water hammer controlling methods, the air vessel protection device is studied widely due to high reliability [9–12]. The pipeline draining or emptying is the cause of some pipe buckling events in the real-world due to the existence of air pockets in the system that expand during the emptying process. Based on the polytropic law, the air pocket expansion leads to a pressure drop in the air pocket, which is capable of creating a sub-atmospheric pressure situation. The emptying process has been studied previously numerically and experimentally [13–15]. Furthermore, recent works have been developed to understand the behavior of an air pocket and setting operational rules in the emptying process to prevent future accidents. Several works have been published recently on the numerical simulation of the emptying process using one-dimensional (1D) models in case of having air inside a pipe [16,17]. Other works using advanced CFD techniques have been undertaken on the dynamic behavior of an air pocket over drainage and the effect of backflow air entrance [15,18]. During a flow establishment in a partially or fully empty pipeline, which is usually referred to as the filling process, an air compression situation may emerge. The filling process is capable of inducing very huge spikes of pressure that can lead to damage to the pipeline equipment or eventually induce a pipe rupture. Previous studies have focused on different aspects of the filling process under different conditions [19–23]. The thermodynamic behavior of the system plays a significant role in the transient phenomena. They can be categorized into slow and fast transient events. During a slow transient phenomenon, the heat transfer might be important, and generally, they obey the polytropic law [24]. Nevertheless, in a fast transient phenomenon, the heat transfer can be neglected, and for some ranges of the flow condition, the polytropic law is not valid [25]. In this context, the current study endeavors to reveal some aspects of the inconsistent behavior of the air pocket during the filling process. More studies are needed to address the protection devices or methods for all mentioned transient situations. Usually, surge tanks, air vessels, and different operational valves are used in pipeline systems to prevent extraordinary pressure magnitudes. Among them, air valves are widely used due to the simple application and reliability. However, air valves can act unexpectedly in some transient scenarios. The air valves have been studied by different authors to understand the application of an air valve and its effect on the response of the system [26–30]. In a recent study [31], a mathematical model was developed, which was appropriately verified by experimental results to study the various aspects of the emptying and filling processes, including the effect of air valves. The mathematical model revealed that increasing the air valve size will reduce the spike of the pressure head for the filling process condition. But this was challenged widely by real-world problems, an issue that shows the deficiency of current mathematical models in the prediction of the variation of parameters for the filling process. This motivated the authors of this paper to study the effect of the air valve on the filling process more deeply using mathematical models and present the main source of the problem in the formulation. This research explains in detail, the mathematical model, including all hydraulic and thermodynamic formulations in comparison to a previous publication [31] and tackles the experimental and numerical study of the filling process with entrapped air when different air release conditions are provided. This current study provides extensive information about the unusual behavior of an air pocket with different sizes of air releasement. Indeed, plenty of mathematical models exist for the filling process, but in general, the behavior of air valves has not been addressed.

2. Mathematical Model

This section presents the mathematical model to simulate a filling process with an air valve in a single pipeline. This process should be a controlled operation, where an entrapped air pocket is compressed by an energy source, and an air valve expels an air volume to relieve pressure surges. Figure 1 shows the scheme of a single pipeline, which consists of a pump or a high-pressure air tank, a length of the filling column, an air valve located at the downstream end, a regulating valve located at the upstream end, and a sloped pipe.

2.1. Assumptions

The mathematical model assumes the uniform movement of the filling water column. The following assumptions are considered:

- The filling water column is modeled using a rigid column model.
- The air–water interface is considered perpendicular to the main direction of a single pipeline.

- The friction factor is constant over the transient event.
- A polytropic model describes the air phase.

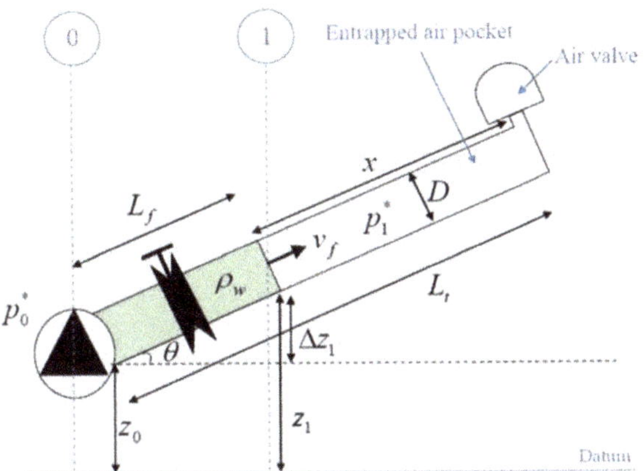

Figure 1. Scheme of a filling process in a single pipeline with an air valve.

2.2. Formulations

Based on the aforementioned assumptions, the filling process can be modeled using the following formulations:

- The mass oscillation equation [1,28,31]: This equation represents the water movement adequately since in transient flows with trapped air, the compressibility of the air is much higher compared to the water and pipe system:

$$\frac{dv_f}{dt} = \frac{p_0^* - p_1^*}{\rho_w L_f} + g\frac{\Delta z_1}{L_f} - f\frac{v_f|v_f|}{2D} - \frac{R_v g A^2 v_f |v_f|}{L_f}, \tag{1}$$

where v_f = water velocity, p_0^* = initial pressure supplied by tank or pump, p_1^* = air pocket pressure, L_f = length of the filling column, t = time, D = internal pipe diameter, R_v = resistance coefficient of the regulating valve, f = friction factor, ρ_w = water density, g = gravity acceleration, Δz_1 = difference elevation, and A = cross-sectional area. The relation $\Delta z_1/L_f$ (named as gravity term) is calculated for single pipelines as $\sin\theta$, where θ represents the pipe slope.

- The air–water interface [28,29]: A piston flow model is considered to represent the interface position, which is applicable in inclined piping installations:

$$\frac{dL_f}{dt} = v_f, \tag{2}$$

- The polytropic model of the air phase [30]: This formulation shows the evolution of the air pocket pressure over time by relating the compression of an air pocket (dV_a/dt) to the quantity of the expelled air by an air valve (dm_a/dt):

$$\frac{dp_1^*}{dt} = k\frac{p_1^*}{V_a}\left(\frac{dV_a}{dt} - \frac{1}{\rho_a}\frac{dm_a}{dt}\right), \tag{3}$$

where k = polytropic coefficient, V_a = air pocket volume, ρ_a = air density, and m_a = air pocket mass.

- The air mass equation [28]:

$$\frac{dm_a}{dt} = -\rho_a v_a A_{exp}, \tag{4}$$

where A_{exp} = cross-sectional area of an air valve for expelling conditions, and v_a = air velocity. Here the air pocket density (ρ_a) inside of a pipe system is identical to the air density expelled by an air valve and considering $m_a = \rho_a V_a$, thus:

$$\frac{dm_a}{dt} = \frac{d(\rho_a V_a)}{dt} = \frac{d\rho_a}{dt}V_a + \frac{dV_a}{dt}\rho_a = -\rho_a v_a A_{exp}. \tag{5}$$

Based on the variables and parameters shown in Figure 2, then:

$$V_a = Ax = A\left(L_t - L_f\right), \tag{6}$$

where x = air pocket size, and L_T = total length of the pipe.

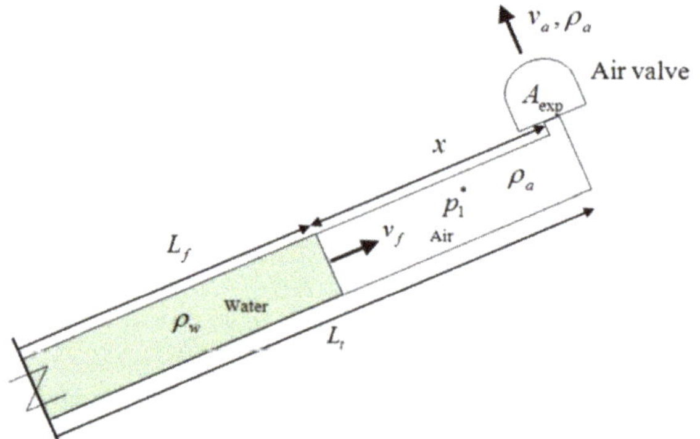

Figure 2. Location of the air valve.

And deriving the Formulation (6), then:

$$\frac{dV_a}{dt} = -Av_f. \tag{7}$$

Plugging Formulations (6) and (7) into (5), then:

$$\frac{d\rho_a}{dt} = \frac{v_f A\rho_a - \rho_a v_a A_{exp}}{A\left(L_t - L_f\right)}, \tag{8}$$

- The air valve characterization: Subsonic conditions are required to perform an adequate filling process according to recommendations given by the American Water Works Association (AWWA) [32], thus:

$$v_a = C_{exp}p_1^* \sqrt{\frac{7}{RT}\left[\left(\frac{p_{atm}^*}{p_1^*}\right)^{1.4286} - \left(\frac{p_{atm}^*}{p_1^*}\right)^{1.714}\right]}, \tag{9}$$

where p_{atm}^* = atmospheric pressure, C_{exp} = outflow discharge coefficient, R = air constant, and T = air temperature.

2.3. System Equations and Resolution

A 5×5 system of algebraic-differential Equations (1)–(3), (8), and (9) describes the filling operation in single pipelines. The system has five unknown hydraulic and thermodynamic variables: v_f, L_f, p_1^*, ρ_a, and v_a. The resolution is conducted using Simulink in MATLAB.

2.4. Initial and Boundary Conditions

The system is considered initially static at $t = 0$. Therefore, the initial conditions are described by $v_f(0) = 0$, $L_f(0) = L_{f,0}$, $p_1^*(0) = p_{1,0}^*$, $\rho_a = 1.205$ kg/m^3, and $v_a(0) = 0$.

3. Numerical Validation

3.1. Experimental Facility and Instrumentation

The mathematical model was validated by experimental tests accomplished in the hydraulic lab of the Instituto Superior Técnico located at the University of Lisbon (Lisbon, Portugal). Sudden pressurization of a trapped air pocket in an undulating pipeline when an air valve was located at the highest point of the pipeline has been addressed. Tests were done in a pipeline having a hydro-pneumatic tank (HT) of 1 m^3 upstream to produce the required initial pressure (p_0^*) for the tests (see Figure 3). For each test, an air pocket was located at the highest point of the pipeline extending towards the upstream branch of the pipeline. An electro-pneumatic ball valve (BV) was used as a means to isolate the pipeline from the high pressure of the HT prior to starting of the test. So, before starting the test, the BV was closed and pressure in the pipeline was set at the atmospheric range. After adjusting the pressure of the HT (using a pressure gauge), the pressure in the pipeline and the air pocket size, the test started by opening the BV. The BV actuation time was 0.20 s leading to sudden pressurization of the downstream pipeline. The pressure data were recorded by a pressure transducer located at Y = 0.8 m and X = 0 m, which had a frequency of data collection of 0.0062 s. The pressure transducer was able to record the absolute pressure up to 25 bar having a maximum pressure measurement error of 0.5% as reported by the manufacturer. This measurement error was negligible compared to the maximum pressure values attained. The pipeline was composed of several polyvinyl chloride (PVC) pipes creating a length of 7.30 m from the HT to the end of the pipeline. The tests were done by changing different parameters, such as the upstream HT pressure and the air pocket size. A commercial air valve S050 (A.R.I. manufacturer) was used in all tests, which had an internal diameter of 3.175 mm ($A_{exp} = 7.92 \times 10^{-6}$ m^2) with an outflow discharge coefficient of 0.32. The resistance coefficient of the BV was 2.2×10^5 ms^2/m^6 for a total opening. The analyzed hydraulic installation can be considered as a single pipeline since transient events occur in the sloped pipe branch due to the valve located at X = 3.0 m remained closed during the experiments.

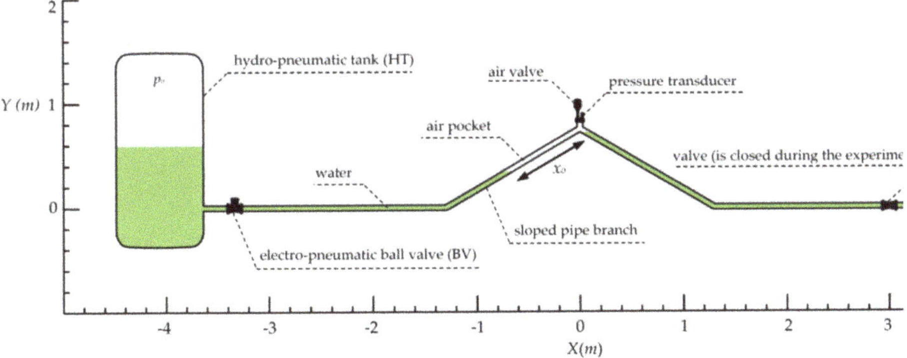

(**a**) Elements and lengths of the experimental apparatus

Figure 3. *Cont.*

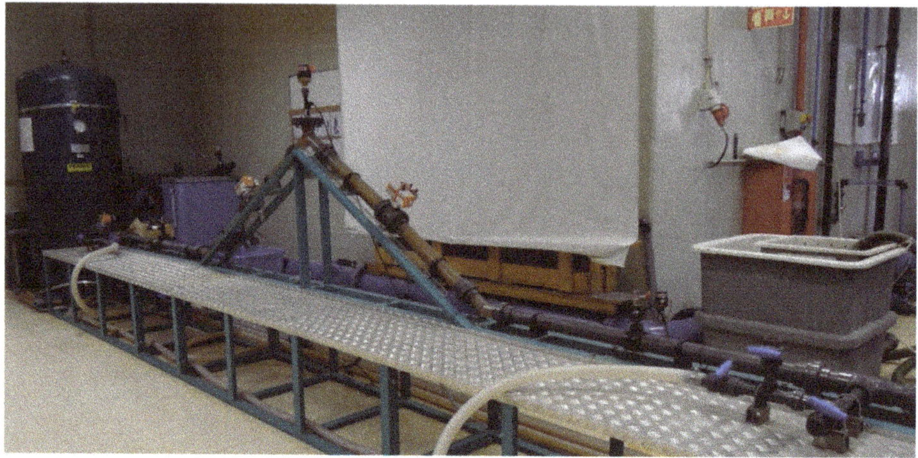

(**b**) Photograph of the experimental apparatus.

Figure 3. Schematic of filling process apparatus.

3.2. Experimental Test

A total of 8 experimental tests were performed to validate the mathematical model proposed by the authors repeating each measurement twice. Initial air pocket sizes (x_0) between 0.96 m and 1.36 m were defined in the 1.50-m-long sloped branch pipe in combination with initial gauge pressures (p_0) of 0.2, 0.5, 0.75, and 1.25 bar in the hydro-pneumatic tank (see Table 1).

Table 1. Characteristics of tests.

Test No.	p_0(bar)	p_0^{*} [1] (Pa)	x_0(m)
1	0.20	120060	0.96
2	0.20	120060	1.36
3	0.50	150075	0.96
4	0.50	150075	1.36
5	0.75	175087	0.96
6	0.75	175087	1.36
7	1.25	225112	0.96
8	1.25	225112	1.36

[1] Absolute pressure in the hydro-pneumatic tank (p_0^{*}) were computed as $p_0 + p_{atm}^{*}$.

3.3. Model Verification

To verify the proposed model, comparisons between computed and measured air pocket pressure oscillations were conducted using a constant friction factor of 0.018, considering the previous work published by the authors [16]. The water column located from X = 0 m to X = 3.4 m (see Figure 3) represents a boundary condition of the system since according to the observations it remains static during all measurements; then, it was neglected for the analysis in the mathematical model. The initial length of the water column was always located at the sloped pipe branch (between X= −1.3 m and X= 0 m). Based on these considerations, the hydraulic installation can be modeled as a single pipeline. For the analyses, the proposed model was developed to simulate the filling process until the closure of the air valve, when a single-phase flow (only water) was reached.

Comparisons show that the mathematical model exhibited a good agreement in following the behavior of the air pocket pressure patterns for the first oscillation compared to the measurements, as shown in Figure 4. However, the mathematical model could not simulate the sub-sequence oscillations because the impact of the water column (from X = −3.2 m to X = 0 m) with the blocking water column (from X = 0 m to 3.4 m) is a complex phenomenon where the air–water interface is not perpendicular to the main direction of the pipe installation.

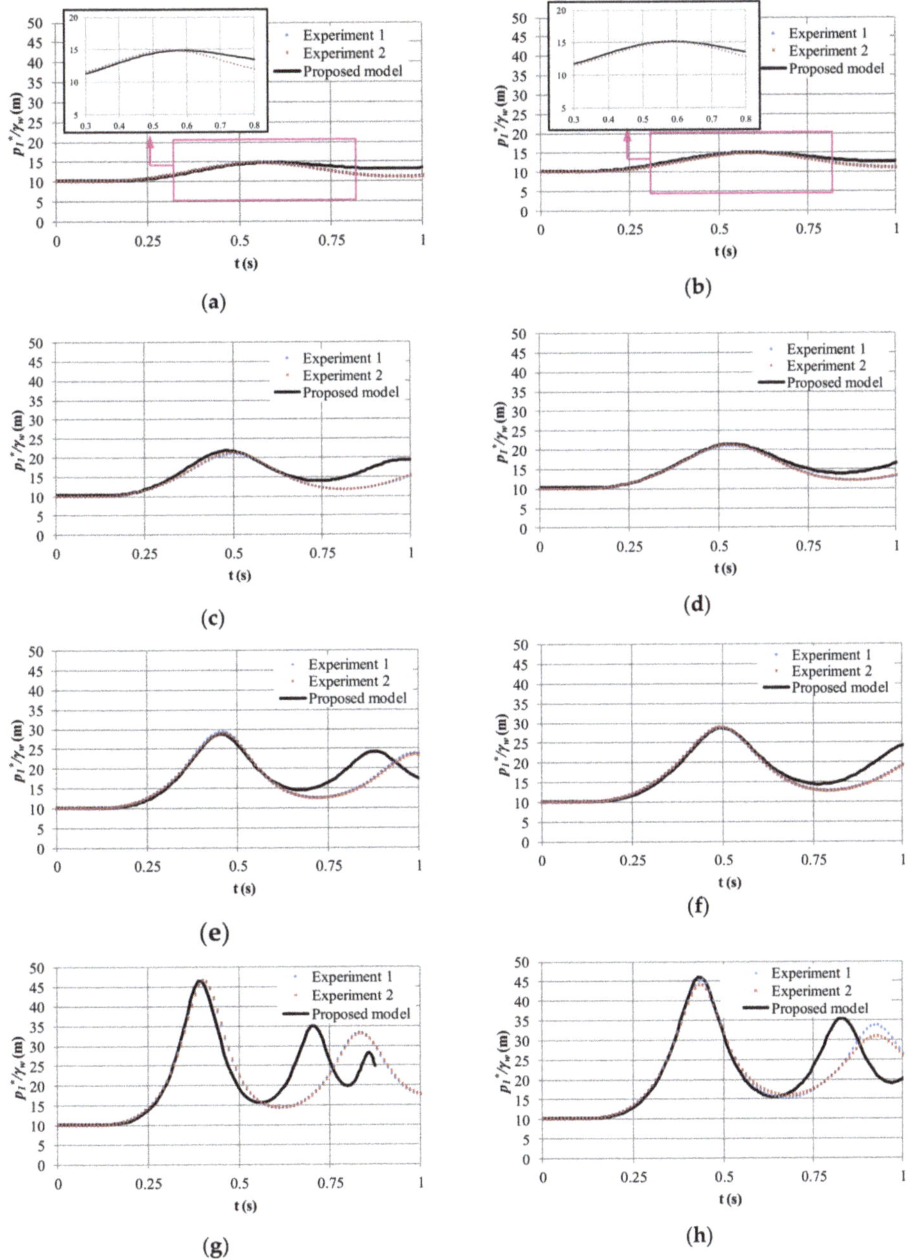

Figure 4. Air pocket pressure patterns: (**a**) Test No. 1; (**b**) Test No. 2; (**c**) Test No. 3; (**d**) Test No. 4; (**e**) Test No. 5; (**f**) Test No. 6; (**g**) Test No. 7; (**h**) Test No. 8.

The more important parameter is the hydro-pneumatic tank pressure since its variation implies important differences of values of air pocket pressures. The greater the hydro-pneumatic tank pressure (p_0^*), the higher the air pocket pressure patterns obtained. With a hydro-pneumatic tank pressure of 0.2 bar (Tests No. 1 and No. 2) a maximum value of air pocket pressure head of 15.0 m was reached; in contrast, using a hydro-pneumatic tank pressure of 1.25 bar, peak values of absolute pressure head of 46.9 m and 44.9 m for Test No. 7 and No. 8, respectively, were reached. Peak values of the air pocket pressure were reached at peak time (t_{peak}). The greater the hydro-pneumatic tank pressure, the lower values of t_{peak} obtained, indicating that a faster compression of the entrapped air pocket was attained. For a hydro-pneumatic tank pressure of 0.5 bar, values of t_{peak} of 0.50 s and 0.52 s for Tests No. 3 and No. 4 were attained, respectively; while for a hydro-pneumatic tank pressure of 0.75 bar, values of t_{peak} of 0.46 s and 0.49 s for Test No. 5 and No. 6 were reached, respectively. The greater the air pocket size (x_0), the higher values of t_{peak} reached.

On the other hand, the polytropic equation is explained with Tests No. 7 and No. 8 (using an HT of 1.25 bar), where air pocket sizes of 0.96 and 1.36 m generated air pocket pressure heads of 46.9 and 44.9 m, respectively. The smaller the air pocket size, the greater peak of pressure surges attained. The remaining tests do not show representative differences on the reached maximum air pocket pressure because the initial hydro-pneumatic tank pressures were not so high as to appreciate these differences. For instance, a peak value of air pocket pressure head of 21.4 m was reached for Tests No. 3 and No. 4 with initial air pocket sizes of $x_0 = 0.96$ m and $x_0 = 1.36$ m, respectively. Both the experiment and the mathematical model present these trends.

A summary of experimental results is presented in Table 2, which shows a comparison between maximum values of air pocket pressure head, air pocket size, and attained t_{peak}.

Table 2. Summary of experimental results.

Test No.	Maximum Value of Air Pocket Pressure Head (m)	x_0(m)	t_{peak} (s)
1	15.0	0.96	0.55
2	15.0	1.36	0.58
3	21.4	0.96	0.50
4	21.4	1.36	0.52
5	29.3	0.96	0.46
6	29.1	1.36	0.49
7	46.9	0.96	0.40
8	44.9	1.36	0.44

The prediction of the mathematical model can be observed in Figure 5. It demonstrates how the mathematical model developed by the authors has a good agreement with the computation of the maximum air pocket pressure when it is compared with measured values. Hence, the mathematical model can be used to compute the maximum values of air pocket pressure during a filling operation in water installation. It is important to note that the mathematical model properly reproduces the first oscillation and the maximum absolute pressure, but it is not valid for the rest of the hydraulic event.

The selection of a pipe class should consider not only pressure surges occurrence caused by a pump's stoppages or rapid closure of valves but also the peak value reached by the compression of an air pocket during a filling operation.

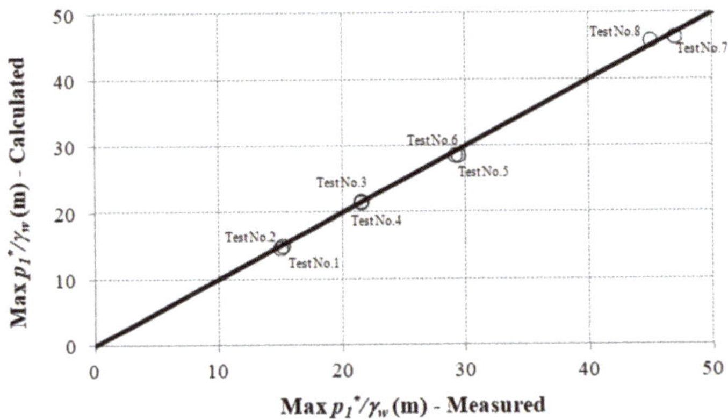

Figure 5. Comparison between calculated and measured maximum air pocket pressure.

3.4. Comparisons Without Air Valve

To note the action of the air valve S050 on the behavior of the air pocket pressure patterns and how this device can relieve pressure surges occurrence, a comparison of results between using the air valve S050 and neglecting it was conducted in the experimental facility. Figure 6 shows the comparison for Test No. 5, where the air pocket pressure pattern exhibited a similar trend for these two scenarios. The mathematical model presented a better behavior in the prediction of absolute pressure oscillations when there was no air valve compared to the scenario using the air valve S050. The prediction of peak values of air pocket pressure head for both scenarios was detected by the mathematical model, where a maximum value of 32.2 m (at 0.44 s) was reached without air valve, and using the air valve S050 the peak value was 29.3 m (at 0.46 s).

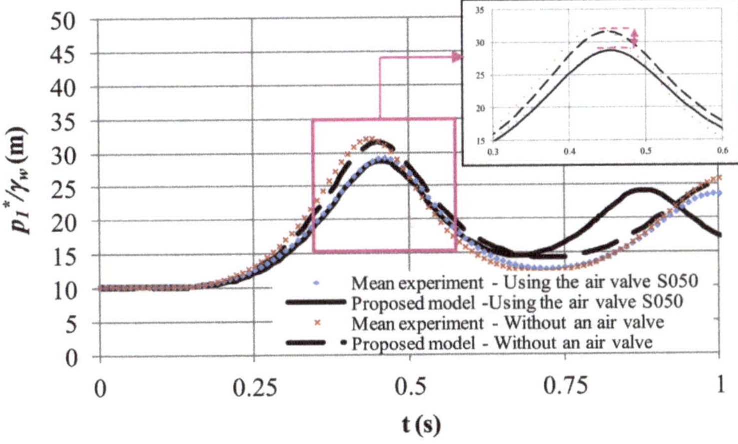

Figure 6. Effect of air pocket pressure pattern considering and neglecting the air valve S050 for Test No. 5.

The air valve S050 is used in hydraulic installations to release air bubbles when pipelines are completely occupied by water under a normal situation of operation. The air valve S050 can be used during filling processes, which can reduce low percentages of the maximum air pocket pressure since its small outlet orifice is 3.175 mm, as mentioned by the manufacturer A.R.I. Figure 7 and Table 3 present the peak values of the air pocket pressure head reached. Results show how the air valve can relieve pressure surges from 5% to 9% compared to the scenario when there was no installed air valve.

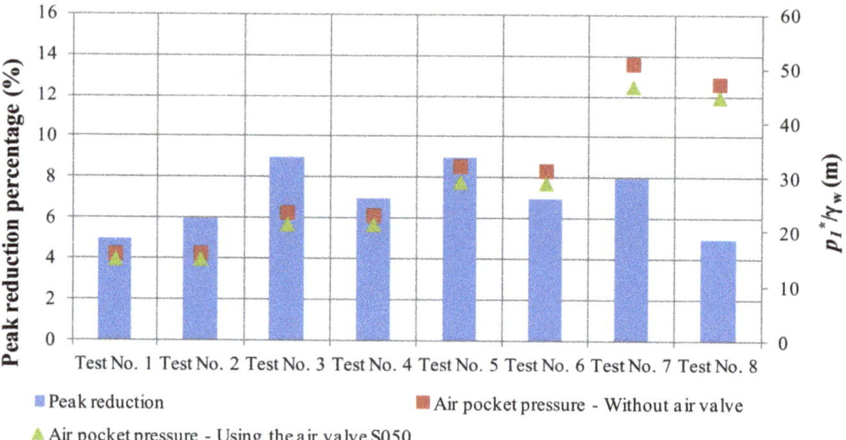

Figure 7. Peak reduction percentage vs. maximum air pocket pressure attained.

Table 3. Summary of extreme values of reached pressure surges.

Test No.	Air Pocket Pressure Head (M)		Peak Reduction Percentage (%)
	Without Air Vale	Using the Air Valve S050	
1	15.9	15.0	5
2	16.0	15.0	6
3	23.6	21.4	9
4	23.1	21.4	7
5	32.2	29.3	9
6	31.4	29.1	7
7	51.1	46.9	8
8	47.4	44.9	5

4. Conclusions

Filling maneuvers in water pipelines generate pressure surges since air pockets are being compressed. The analysis of filling processes without air valves has been studied in detail in recent years; however, there are few studies related to the effects of air valves on the upsurge control, which need a better understanding to reduce pipeline failures during these processes. Air valves need to be positioned along hydraulic installations to expel enough volume of air to relieve peak values induced by air pocket compression.

This research presented a 1D mathematical model to simulate the hydraulic behavior of a water column and the thermodynamic evolution of an air pocket during a filling operation using a commercial air valve. The mathematical model was validated in an experimental facility composed by a 7.3-m-long PVC pipeline with an internal diameter of 63 mm. Air pocket pressure patterns were measured for eight different tests. Comparisons of the air pocket pressure between computed and measured values indicated how the mathematical model is suitable to predict the behavior of the first oscillation, which is very important considering the extreme values of absolute pressure are attained in this period. However, the mathematical model is not valid for the rest of the transient response since the impact between the water column and the blocking water column produces a complex phenomenon, where the air–water interaction is not perpendicular to the main direction of the water pipeline.

The air valve S050 relieves the peaks of air pocket pressure in a ratio ranging from 5% to 9% in the laboratory pipe scale compared to the scenario when this device was not installed. The relief percentages of the maximum air pocket pressure were obtained because the air valve S050 presents a small outlet orifice of 3.175 mm.

The mathematical model can be used to compute the maximum air pocket pressure during filling processes using both undersized and well-sized air valves. However, the analysis of oversized air valves was not covered using the mentioned formulations since more extreme pressure surges can be achieved depending on air pocket size and initial hydro-pneumatic tank pressure.

Author Contributions: Conceptualization, Ó.E.C.-H. and V.S.F.-M.; Data curation, Ó.E.C.-H. and M.B.; Methodology, V.S.F.-M.; Writing—original draft, Ó.E.C.-H. and M. B.; Writing—review and editing, H.M.R.

Funding: This work is supported by Fundacao para a Ciencia e Tecnologia (FCT), Portugal (grant number PD/BD/114459/2016).

Conflicts of Interest: The authors declare no conflict of interest.

Abbreviations

The following abbreviations are used in this manuscript:

A	Cross-sectional area of pipe (m^2)
A_{exp}	Cross-sectional area of outlet orifice in an air valve (m^2)
C_{exp}	Outflow discharge coefficient an air valve ($-$)
D	Internal pipe diameter (m)
f	Friction factor ($-$)
k	Polytropic coefficient ($-$)
g	Gravity acceleration (m/s^2)
L_f	Length of the filling column (m)
L_t	Total length of pipe (m)
m_a	Air mass (kg)
p^*_{atm}	Atmospheric pressure (Pa)
p^*_0	Absolute pressure supplied by an energy source (Pa)
p^*_1	Air pocket pressure (Pa)
R	Gas constant (287 J/kg/ K)
R_v	Resistance coefficient of the regulating valve (m s^2/m^6)
T	Air temperature (K)
t	Time (s)
t_{peak}	Peak time (s)
V_a	Air volume (m^3)
v_a	Air velocity (m/s)
v_f	Water velocity (m/s)
x	Air pocket size (m)
Δz_1	Difference elevation (m)
ρ_a	Air density (kg/m^3)
ρ_w	Water density (kg/m^3)
BV	Electro-pneumatic ball valve
HT	Hydro-pneumatic tank

References

1. Abreu, J.; Cabrera, E.; Izquierdo, J.; García-Serra, J. Flow Modeling in Pressurized Systems Revisited. *J. Hydraul. Eng.* **1999**, *125*, 1154–1169. [CrossRef]
2. Azoury, P.H.; Baasiri, M.; Najm, H. Effect of Valve-Closure Schedule on Water Hammer. *J. Hydraul. Eng.* **1986**, *112*, 890–903. [CrossRef]
3. Himr, D. Investigation and Numerical Simulation of a Water Hammer with Column Separation. *J. Hydraul. Eng.* **2015**, *141*, 04014080. [CrossRef]

4. Simpson, A.R.; Wylie, E.B. Large Water-Hammer Pressures for Column Separation in Pipelines. *J. Hydraul. Eng.* **1991**, *117*, 1310–1316. [CrossRef]

5. Saemi, S.; Raisee, M.; Cervantes, M.J.; Nourbakhsh, A. Computation of two- and three-dimensional water hammer flows. *J. Hydraul. Res.* **2019**, *57*, 386–404. [CrossRef]

6. Karney, B.W.; Simpson, A.R. In-line check valves for water hammer control. *J. Hydraul. Res.* **2007**, *45*, 547–554. [CrossRef]

7. Triki, A. Water-Hammer Control in Pressurized-Pipe Flow Using a Branched Polymeric Penstock. *J. Pipeline Syst. Eng. Pract.* **2017**, *8*, 04017024. [CrossRef]

8. Triki, A.; Fersi, M. Further investigation on the water-hammer control branching strategy in pressurized steel-piping systems. *Int. J. Press. Vessel. Pip.* **2018**, *165*, 135–144. [CrossRef]

9. Stephenson, D. Simple Guide for Design of Air Vessels for Water Hammer Protection of Pumping Lines. *J. Hydraul. Eng.* **2002**, *128*, 792–797. [CrossRef]

10. Besharat, M.; Tarinejad, R.; Ramos, H.M. The Effect of Water Hammer on a Confined Air Pocket Towards Flow Energy Storage System. *J. Water Supply: Res. Technol.-AQUA.* **2016**, *65*, 116–126. [CrossRef]

11. Besharat, M.; Tarinejad, R.; Aalami, M.T.; Ramos, H.M. Study of a Compressed Air Vessel for Controlling the Pressure Surge in Water Networks: CFD and Experimental Analysis. *Water Resour. Manag.* **2016**, *30*, 2687–2702. [CrossRef]

12. Besharat, M.; Viseu, M.T.; Ramos, H.M. Experimental Study of Air Vessel Sizing to either Store Energy or Protect the System in the Water Hammer Occurrence. *Water* **2017**, *9*, 63. [CrossRef]

13. Laanearu, J.; Annus, I.; Koppel, T.; Bergant, A.; Vučkovič, S.; Hou, Q.; van't Westende, J.M.C. Emptying of Large-Scale Pipeline by Pressurized Air. *J. Hydraul. Eng.* **2012**, *138*, 1090–1100. [CrossRef]

14. Tijsseling, A.; Hou, Q.; Bozkus, Z.; Laanearu, J. Improved One-Dimensional Models for Rapid Emptying and Filling of Pipelines. *J. Press. Vessel Technol.* **2016**, *138*, 031301. [CrossRef]

15. Besharat, M.; Coronado-Hernández, O.E.; Fuertes-Miquel, V.S.; Viseu, M.T.; Ramos, H.M. Backflow Air and Pressure Analysis in Emptying Pipeline Containing Entrapped Air Pocket. *Urban Water J.* **2018**, *15*, 769–779. [CrossRef]

16. Coronado-Hernández, O.E.; Fuertes-Miquel, V.S.; Besharat, M.; Ramos, H.M. Subatmospheric Pressure in a Water Draining Pipeline with an Air Pocket. *Urban Water J.* **2018**, *15*, 346–352. [CrossRef]

17. Coronado-Hernández, O.E.; Fuertes-Miquel, V.S.; Iglesias-Rey, P.L.; Martínez-Solano, F.J. Rigid Water Column Model for Simulating the Emptying Process in a Pipeline Using Pressurized Air. *J. Hydraul. Eng.* **2018**, *144*, 06018004. [CrossRef]

18. Besharat, M.; Coronado-Hernández, O.E.; Fuertes-Miquel, V.S.; Viseu, M.T.; Ramos, H.M. Computational Fluid Dynamics for Sub-Atmospheric Pressure Analysis in Pipe Drainage. *J. Hydraul. Res.* **2019**, 1–13. [CrossRef]

19. Vasconcelos, J.G.; Wright, S.J. Rapid Flow Startup in Filled Horizontal Pipelines. *J. Hydraul. Eng.* **2008**, *134*, 984–992. [CrossRef]

20. Trindade, B.C.; Vasconcelos, J.G. Modeling of Water Pipeline Filling Events Accounting for Air Phase Interactions. *J. Hydraul. Eng.* **2013**, *139*, 921–934. [CrossRef]

21. Malekpour, A.; Karney, B.; Nault, J. Physical understanding of sudden pressurization of pipe systems with entrapped air: Energy auditing approach. *J. Hydraul. Eng.* **2015**, *142*, 04015044. [CrossRef]

22. Apollonio, C.; Balacco, G.; Fontana, N.; Giugni, M.; Marini, G.; Piccinni, A.F. Hydraulic Transients Caused by Air Expulsion during Rapid Filling of Undulating Pipelines. *Water* **2016**, *8*, 25. [CrossRef]

23. Wang, L.; Wang, F.; Karney, B.; Malekpour, A. Numerical Investigation of Rapid Filling in Bypass Pipelines. *J. Hydraul. Res.* **2017**, *55*, 647–656. [CrossRef]

24. Chaudhry, M.H. *Applied Hydraulic Transients*, 3rd ed.; Springer: New York, NY, USA, 2014.

25. Besharat, M.; Coronado-Hernández, O.E.; Fuertes-Miquel, V.S.; Viseu, M.T.; Ramos, H.M. CFD and 1D Simulation of Transient Flow Effect on Air Vessel. In Proceedings of the 13th International Conference on Pressure Surges, Bordeaux, France, 14–16 November 2018; BHR Group: Bordeaux, France, 2018.

26. Ramezani, L.; Karney, B.; Malekpour, A. The Challenge of Air Valves: A Selective Critical Literature Review. *J. Water Resour. Plan. Manag.* **2016**, *141*, 04015017. [CrossRef]

27. Balacco, G.; Apollonio, C.; Piccinni, A.F. Experimental Analysis of Air Valve Behaviour During Hydraulic Transients. *J. Appl. Water Eng. Res.* **2015**, *3*, 3–11. [CrossRef]

28. Fuertes-Miquel, V.S.; López-Jiménez, P.A.; Martínez-Solano, F.J.; López-Patiño, G. Numerical modelling of pipelines with air pockets and air valves. *Can. J. Civ. Eng.* **2016**, *43*, 1052–1061. [CrossRef]
29. Fuertes-Miquel, V.S.; Coronado-Hernández, O.E.; Iglesias-Rey, P.L.; Mora-Melia, D. Transient Phenomena during the Emptying Process of a Single Pipe with Water-Air Interaction. *J. Hydraul. Res.* **2019**, *57*, 318–326. [CrossRef]
30. Coronado-Hernández, O.E.; Fuertes-Miquel, V.S.; Besharat, M.; Ramos, H.M. Experimental and Numerical Analysis of a Water Emptying Pipeline Using Different Air Valves. *Water* **2017**, *9*, 98. [CrossRef]
31. Coronado-Hernández, O.E.; Fuertes-Miquel, V.S.; Besharat, M.; Ramos, H.M. A Parametric Sensitivity Analysis of Numerically Modelled Piston-Type Filling and Emptying of an Inclined Pipeline with an Air Valve. In Proceedings of the 13th International Conference on Pressure Surges, Bordeaux, France, 14–16 November 2018; BHR Group: Bordeaux, France, 2018.
32. American Water Works Association (AWWA). *Manual of Water Supply Practices-M51: Air-Release, Air-Vacuum, and Combination Air Valves*, 1st ed.; American Water Works Association: Denver, CO, USA, 2001.

© 2019 by the authors. Licensee MDPI, Basel, Switzerland. This article is an open access article distributed under the terms and conditions of the Creative Commons Attribution (CC BY) license (http://creativecommons.org/licenses/by/4.0/).

Article

A New Technical Concept for Water Management and Possible Uses in Future Water Systems

Pål-Tore Storli [1,*] **and T. Staffan Lundström** [2]

[1] Department of Energy and Process Engineering, Faculty of Engineering Sciences, Norwegian University of Science and Technology, 7491 Trondheim, Norway

[2] Division of Fluid and Experimental Mechanics, Luleå University of Technology, 97187 Luleå, Sweden; staffan.lundstrom@ltu.se

* Correspondence: pal-tore.storli@ntnu.no; Tel.: +47-9778-2146

Received: 26 October 2019; Accepted: 28 November 2019; Published: 29 November 2019

Abstract: A new degree of freedom in water management is presented here. This is obtained by displacing water, and in this paper is conceptually explained by two methods: using an excavated cavern as a container for compressed air to displace water, and using inflatable balloons. The concepts might have a large impact on a variety of water management applications, ranging from mitigating discharge fluctuation in rivers to flood control, energy storage applications and disease-reduction measures. Currently at a low technological readiness level, the concepts require further research and development, but the authors see no technical challenges related to these concepts. The reader is encouraged to use the ideas within this paper to find new applications and to continue the out-of-the-box thinking initiated by the ideas presented in this paper.

Keywords: water management; reservoirs; hydropower plants; pumped storage power plants; hydropeaking; environmental flows

---

1. Introduction

Fresh water is a paramount part of any human life and a prerequisite for a developed society and high quality of life. Apart from the obvious use of drinking it, water can provide many services for both society and environments, either in a natural or modified system. The IPCC 2011 climate report (chapter five) has an extensive section (on services and characteristics of Hydro Power Plant (HPP) projects, but transferrable to the general case as well) in which energy and water management services as well as their main environmental and societal characteristics are identified.

For society, these services might be the following: transportation and recreation (riverboats and barges); irrigation; fisheries; energy storage and power production (hydropower production (at HPP)); electric energy storage (Pumped Hydro Storage, (PHS)); waste management; fertilization of farmland (sedimentation by flooding); protection of shorelines (sediments settling in estuaries and river mouths dampening waves and tidal erosion); flood protection (dams and natural lakes reduce flood intensity); and groundwater stabilization.

For environments, the services might be as follows: providing habitat and reproduction areas for living organisms ranging from the smallest protozoa and algae to plants, fish, birds and mammals; the cleaning of river beds during floods; bringing nutrients in the form of sediments to coastal regions; and the settling of sediments to protect ocean shorelines and thus habitats.

The availability of fresh water varies between different parts of the world. Consequently, the use and management of water also vary between countries and continents. Climate changes are predicted to result in changing weather systems, altering the intensity and distribution of precipitation around the globe. The effect of this on the watercourses on a regional and local scale is difficult to predict, as global models must be downscaled into climatic data for use in models of catchments areas [1].

An overall trend seems to be captured in Figure 1, however, which represents the median of 12 climate model projections using the A1B scenario, "A balanced emphasis on all energy sources", from the Special Report on Emissions Scenario by the Intergovernmental Panel on Climate Change (IPCC) [2]. "Climate change has large direct impacts on several aspects of the hydrological cycle—in particular, increases in extremes—making managing and using water resources more challenging" [3].

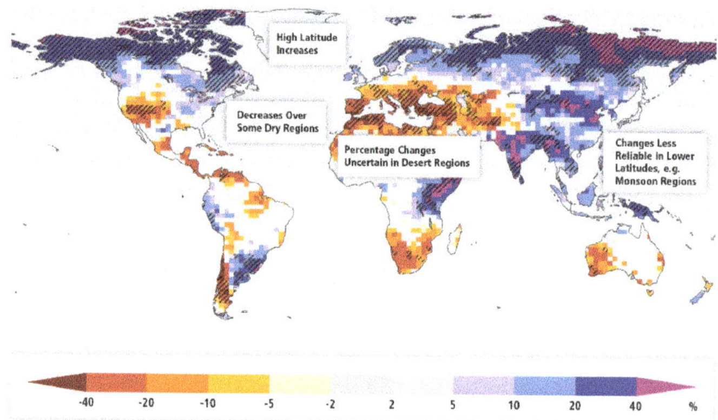

Figure 1. Large-scale changes in annual runoff (water availability, in percent), reprinted from [1].

Current water management systems and infrastructures have been established based on historical data, typically before anthropogenic climate changes were even theorized as being possible. The systems represent large investments and, quite often, incremental additions and/or modifications are not possible, or at least are very expensive. Utilizing the infrastructure to the fullest or even increasing their capacity without modifying the systems is important. When developing the ideas presented in this paper, reusing existing infrastructure has been a guiding principle. The applications presented here are at a low technological readiness level (TRL). As such, cost has not been given any weight in this paper, although it is likely that cost is strongly related to any installation derived from the ideas presented. Hence, a cost–benefit analysis is of course necessary in the long term, but if the benefits are large enough, funding is often available. To put things into perspective, following the 2008 financial crisis, the aid to the financial sector from EU member states was 1400 billion US$ [4], whereas the worldwide government spending on low-carbon Research and Development (R&D) in 2017 was just above 20 billion US$ [5]. This means that close to 70 years of global low-carbon R&D funding was used to keep the EU financial system afloat. This strongly indicates that, where there is a governmental and political will, funding is available.

This paper will present novel ideas on how to achieve a new degree of freedom for topics related to water management. This freedom is linked to the active control of the displacement of water by compressed air in different configurations. The objective of this paper is to present the concepts and to get engineers and scientists to "think outside the box", and in this way stimulate ingenuity and the emergence of new technical solutions. The motivation of the authors is stimulate the development of new technical solutions that will provide water management services to society in the future.

The authors acknowledge that part of this paper has the characteristics of a popular scientific publication more than a technical science publication. However, to be able to bring the concepts in focus to a higher TRL, they need to be presented to a technical audience. Furthermore, the actual ways of implementing the concepts might vary depending on the needs of the services required, so a high degree of tailor-made solutions are likely to be needed. This also suggests that presenting something with a high level of technical detail will not aid the objective of this paper, namely to present the concept and induce creativity within the technical community.

2. New Technical Concepts; Method

The new concepts are based on displacement of water as a method to obtain a new degree of freedom to operate and manage water bodies and flows. Both concepts make use of air as the medium to displace water, but the configurations for obtaining the displacements are different. Problems related to not being able to provide the new degree of freedom are numerous and will be presented in more detail in later sections which will more specifically explain the problems and how displacement can resolve them.

In the next section, the proof for displacement as a method for obtaining a new level of freedom is presented based on first principles. If this proof is not needed from the reader's point of view, this section can be omitted with no loss of continuity.

2.1. Displacing Water Using Compressed Air

Water is, for all practical purposes, incompressible, meaning that a given mass will have a fixed volume. Decoupling water volume and levels means that a portion of the volume must be moved to a different location. This is called displacement and is what the compressed air is claimed to be able to do in the concepts presented.

Although rather intuitive, in the end, it can be shown mathematically to be true as well. This can be done by utilization of the Reynolds Transport Theorem. This theorem can be applied to any extensive property, such as energy, momentum and mass. It basically describes a control volume (CV) which has inlets and outlets crossing a control surface (CS) and formulates that the rate at which an extensive property B_{sys} is produced/consumed in the system must be equal to the sum of the net flow of B through the CS, and the change of the amount of B within the CV according to [6]

$$\frac{dB_{sys}}{dt} = \frac{d}{dt} \int_{CV} \rho b \, dV + \int_{CS} \rho b (\vec{V_r} \cdot \vec{n}) \, dA, \tag{1}$$

where ρ (kg/m^3) is the density of the fluid carrying the property B; $b = B/m$ (B/kg) is the intensive property linked to the extensive property; and $\vec{V_r}$ (m/s) and $\vec{n}$ (-) are the relative velocity vector between the fluid and the local CS and the normal vector of the local CS, respectively. Consider two such systems, one for air (subscript a) contained within another one for water (subscript w), as shown in Figure 2:

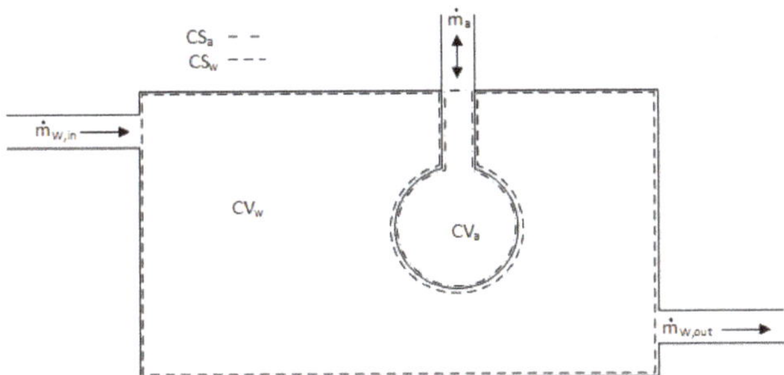

Figure 2. Description of combined air and water systems.

The interphase between the water and air is intended to be flexible—like a balloon—so that it is free to expand or collapse. The quantities $\dot{m}_{w,in}$ and $\dot{m}_{w,out}$ are the mass flow of water in and out of the CV, respectively, while $\dot{m}_a$ is the mass flow of air (positive or negative) through the air inlet.

Two versions of Equation (1) can be constructed, considering that no air or water is produced or consumed ($\frac{DB_a}{Dt} = 0$; $\frac{DB_w}{Dt} = 0$); using $B_1 = m_a$; $b_1 = m_a/m_a = 1$; $B_2 = m_w$; $b_2 = m_w/m_w = 1$;

$$\frac{d}{dt} \int_{CV,w} \rho_w \, dV = - \int_{CS,w} \rho_w (\vec{V_r} \cdot \vec{n}) \, dA = \dot{m}_{w,in} - \dot{m}_{w,out}, \tag{2}$$

$$\tfrac{d}{dt}\int_{CV,a}\rho_a dV = -\int_{CS,a}\rho_a(\vec{V_r}\cdot\vec{n})dA = \dot{m}_a. \tag{3}$$

Assuming the densities to be constant (which holds for low-pressure water, and is a slight simplification for low-pressure air), the densities can be removed in the respective equations, and adding the resulting equations together yields

$$\tfrac{d}{dt}\left(\int_{CV,a} dV + \int_{CV,w} dV\right) = Q_a + Q_{w,in} - Q_{w,out}, \tag{4}$$

where Q is the volume flow (or discharge) of water or air, corresponding to the mass flows. The left-hand side of Equation (4) describes the rate of change of the combined volume of air and water. The right-hand side considers the sum of the discharge through the inlets and outlets. Imagine that a volume of air is submerged in a balloon in a water reservoir (as shown in Figure 2) and for some reason the water level in the reservoir is not allowed to change. This means that the combined volume should be constant, and Equation (4) is simplified to

$$Q_a = Q_{w,out} - Q_{w,in}. \tag{5}$$

If no air volume exists, as is the case for existing reservoirs, Equation (5) reduces to $Q_{w,out} = Q_{w,in}$, meaning that the discharge of water into and out of the reservoir must be identical. Thus, Equation (5) show that the presence of the air displacement makes it possible to decouple the water out of and into the reservoir, obtaining the new degree of freedom.

Energy Considerations

Even if a new degree of freedom is obtained, some insight into the energy cost of this is appropriate. Using energy E (kJ) as the extensive property under investigation in Equation (1), a starting point for an energy analysis of the system can be defined. After divisions of energy into heat (which will be neglected) and work, and further dividing work into work added to the system by a component and pressure work added by the surroundings on the inlets and outlets [6], we get

$$\dot{W}_{shaft} = \tfrac{d}{dt}\int_{CV}\rho e dV + \int_{CS}\rho\left(\tfrac{P}{\rho}+e\right)(\vec{V_r}\cdot\vec{n})dA, \tag{6}$$

where $\dot{W}_{shaft}$ (kJ/s) is the power added or extracted by the component(s); $e = u + \tfrac{V^2}{2} + gz$, where the right-hand side terms are internal, kinetic and potential specific energies, respectively (kJ/kg). This system can be set-up for the water and air configuration in Figure 2, considering the case where the water level must be maintained at a constant level while providing environmental flow $Q_{w,out}$ while there is no inflow, $Q_{w,in} = 0$. The water system has no component that can add (pump) or extract (turbine) power to or from the flow, so there in no $\dot{W}_{shaft}$-term in this system. However, for the air system, a term $\dot{W}_{shaft,a}$ appears to represent the power added to inflate it. The development of the equations can be found in Appendix A, yielding

$$\dot{W}_{shaft,a} = g(\rho_w - \rho_a)(z_s - z_{cg,a})Q_{w,out}, \tag{7}$$

where z_s is the elevation of the reservoir surface and $z_{cg,\,a}$ is the elevation of the centre of gravity (COG) of the air body. Further realising that the density of air is much lower than the water density, the term related to the air system will in practise be negligible; consequently,

$$\dot{W}_{shaft,a} = g\rho_w(z_s - z_{cg,a})Q_{w,out}. \tag{8}$$

This is the power needed to provide the necessary displacement of water by the air system. Scrutinizing Equation (8), it is recognized that the term represents the power needed to bring the flow from atmospheric pressure to the pressure at the COG of the air body. Another case can be investigated by setting the discharge $Q_{w,out}$ to a negative value. This case is one in which the water level must be kept constant, and there is an inflow (negative outflow), but no outflow. In this case, the term $\dot{W}_{shaft,a}$ will change sign, indicating power will go in the opposite direction. This means that regeneration of power is possible when the air is released from the system. For a full cycle, we thus only must pay the price of energy losses, because the inflation/deflating process is an energy storage process.

2.2. Examples of Use of the New Capabilities

There are several examples that can be used to highlight some uses of the new capabilities obtained by this decoupling. First, we assume that the water level at its maximum in a reservoir used for hydropower, for instance. The combined volume cannot be increased, because water would be spilled over the dam or through spillways, and it can be assumed that this must be avoided, perhaps due to safety issues or economical concerns. If the draining of water, $\dot{V}_{w,out}$, due to capacity issues is smaller than the inflow of water, $\dot{V}_{w,in}$, releasing air from the balloon (having negative $\dot{V}_a$) will keep the volume constant. The volume of air in the CV will decrease, allowing for the water volume to increase, but keeping the sum of the two fixed. For this to be possible, air must be present in the volume prior to this incident occurring. However, the reservoir levels and weather forecast should give an indication of such an incident occurring, and the draining of water can be kept higher than the inflow while air is added to the volume in the hours before the incident. Alternatively, some air volume could be present as a default safety margin. If such air volume—in this case, in the form of a balloon—does not exist, there is no possibility at all of avoiding water to be spilled, and safety is no longer ensured.

The opposite case is if the water is at the minimum allowed level; for example, near the end of the dry season. If the need for water downstream supplied by the draining of water, $\dot{V}_{w,out}$, is larger than the inflow of water, $\dot{V}_{w,in}$, this is made possible by adding air to the balloon, having a positive $\dot{V}_a$. Then, the inflating balloon would displace water volume as the water was drained and the level would be kept constant. If such a balloon did not exist, the needed water could not be supplied because this would cause the level to go below the minimum allowable level. Supplying the needed water can, for instance, be of paramount importance to avoid the stranding of fish, [7]. Keeping the level constant can also be of importance for benthic fauna [8].

Both cases above assume that the combined volume should be constant, but the flexibility provided by the displacement can also be used to increase or reduce the combined volume as well, effectively changing the water levels more than what would be the case for mass balance in and out of the reservoirs. Such applications will be discussed later in the paper.

Obtaining the desired displacement can be done in several ways. In principle, a crane could be used to submerge a solid body with a density higher than the water and thus displace a water volume corresponding to the volume of the body. With no indications of these being the only appropriate ways of achieving displacement, two other configurations are considered in this paper; they use an underground cavern and inflatable balloons, respectively. The reasons for choosing these are the fact that much of the infrastructure and components needed are available from other applications existing today and that a gas—in this case, air—is relatively easy to handle. This will be explained in the sections describing the two configurations.

2.2.1. Inflatable Balloons

Displacing water by inflating balloons anchored to the bottom of the water body is the first configuration discussed in this paper. It can be seen in Figure 3.

The principle is intuitively understood and identical to the example used to prove the concept in the section above. Using compressed air to inflate the balloons will displace water, giving the flexibility to operate with a new degree of freedom. Using balloons, the need arises to have strong foundation for the balloons as they experience large buoyancy forces. Although used for a slightly different purpose (compressed air energy storage at great ocean depths), balloons of the type considered appropriate for this configuration have been made and tested in a laboratory [9].

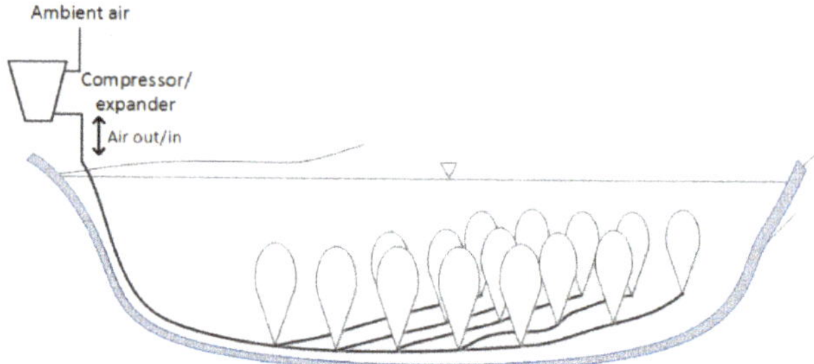

Figure 3. Principal drawing of using inflatable balloons.

2.2.2. Air Cushion Underground Cavern (ACUR)

ACUR is an underground cavern excavated for retaining air near a water surface. It can be seen sketched in Figure 4.

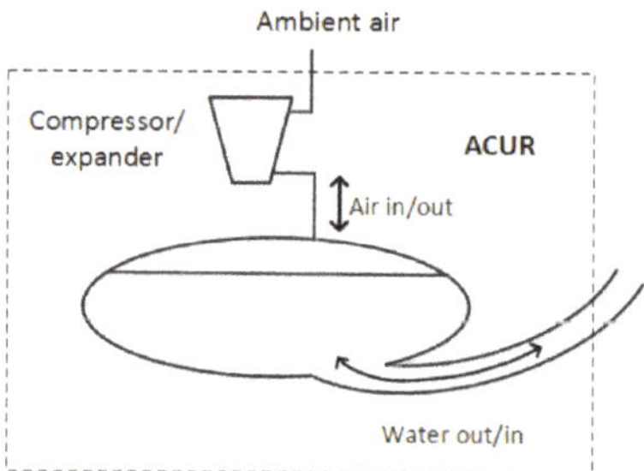

Figure 4. Principal drawing of the Air Cushion Underground Cavern (ACUR).

The tunnel allowing water to go in or out of the ACUR element is connected to an adjacent water reservoir. Displacing water in ACUR by admitting compressed air will give an addition to the water volume of the adjacent reservoir; vice versa, expanding air from ACUR will allow for water entering it, removing the same water volume from the adjacent water body. ACUR is simply an additional volume in which water can be stored. However, the flexibility obtained by using air to displace water is what gives the possibility of the manipulation of water levels.

Subject to acceptable ground conditions, making ACUR water and air-tight might be a challenge. ACUR must be as close to the water level of the adjacent water body as possible, and this will make the pressure inside ACUR quite low—not significantly higher that atmospheric pressure. There is a resemblance between ACUR and a high-pressure component present in a hydropower plant, called air cushion surge chambers. They are chambers in the rock hydraulically connected to the penstock of power plants and are used where a surge shaft cannot be installed due to the local geographical conditions. They are partially filled with water and air, and the purpose of the surge chamber is to improve the governing stability and reduce retardation pressures when stopping the flow through the power plant, either for regulation of power or shutting down the power plant. The fact that these installations can provide sufficient air and water-tightness despite high pressures indicates that it should be possible to make ACUR air and water-tight as well.

3. Applications

There are different applications that can be imagined for these concepts. The balloons clearly need a water volume of some size to work, since they are positioned within the water volume. ACUR provides an additional hidden volume and can be constructed adjacent to small rivers and where no significant volume is present, as well as connected to reservoirs. In addition to flow in regulated rivers, as presented below, the configurations may be used in cities where flooding is a main issue [10]. Watersheds such as rivers, channels, lakes, pools and ponds may be equipped with balloons as shown in Figure 3 to keep the water surfaces at a proper level. During heavy rains, air is released from the balloons, creating a large volume for the water to fill. Once the weather conditions become stable, air can be pumped into the balloons at a slow speed, letting the water leave the city in a controlled fashion.

3.1. Floods, Droughts and Discharge Fluctuation Manipulations

As Figure 1 shows, run-off is expected to change significantly for most of the planet in the future. Floods and droughts are thus likely to intensify. Since simulations have shown that the ACUR configuration can be used to both mitigate and mimic floods [11], it has a potential to be implemented for rivers and used to dampen the magnitude of the discharge in flood incidents. Destructive floods are ones that are so big that they rarely occur, named by the amount of years between their statistical appearance. A 50-year flood is thus more severe than a 10-year flood. Furthermore, it is the flood peak that is responsible for the level of destruction. Hence, by allowing some of the discharge to enter ACUR, the peak flood discharges can be reduced, making the flood less dangerous. Such a scenario is illustrated in Figure 5.

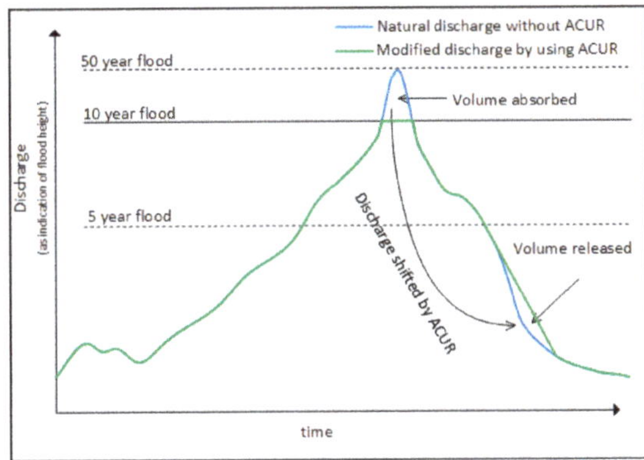

Figure 5. Flood discharge manipulation by ACUR.

The volume needed to turn a 50-year flood into a 10-year flood is of course dependent on the flood characteristics. The duration of the flood peaks tends to be short, so it is not far-fetched to imagine a feasible volume for ACUR being able to relieve the situation. Furthermore, the periods of a flood incident in which the water retrieves rapidly might also be a concern, because an overly high hydro morphological pressure in river embankments might cause landslides. The water stored at the peak flood event could be stored and released to prevent water levels falling too rapidly, hence reducing the risk of landslides. In rivers with many tributaries, ACUR can be used to shift the time of the occurrence of the peak flow of the individual tributaries so that the flood peaks from the tributaries do not all add to the main river flood peak.

In periods of drought, stored water in ACUR can be used for periods where a river flow is at critical levels. The water in the cavern would be subjected to little evaporation, as it is kept away from solar radiation and winds.

Some natural phenomena are triggered by flood-like discharge events. This is might be migratory events for fish in rivers. As an example, the Atlantic salmon migrate from the river as salmon fry in the springtime snow-melt floods [12]. At the other end of the salmon life cycle, the salmon run up the river at high-discharge periods to easily get to the spawning grounds. Mimicking local flood-like events is, for instance, a perfect method to supply water to channels used to attract fish to fish-ways [13,14]. Active use of ACUR could thus facilitate these natural migrations.

In heavily regulated rivers, natural floods have been reduced and silt and sediments have settled in the gravel, which are used by fish as a substrate for hiding fertilized eggs. This has been found to be the case for white sturgeon in the Kootenai River downstream Libby Dam, where the relative rarity of very high flows due to the regulation by the dam is partially responsible for the lack of successful spawning [15]. Therefore, there might be cases in which a regulated river would benefit from a flood-like discharge event. ACUR could be used to provide this as well, by storing water and releasing it when the natural river flow is high, artificially creating a flood-like peak in the discharge. Simply releasing additional high-value water from the reservoir in such a case can be an alternative but carries a high cost. The water stored in ACUR has already given its energy to the generators of the hydroelectric plant and thus has a lower value. The cost of pushing water out of the cavern and into the river would be the cost of operating the compressors providing the pressurized air. In fact, in electrical sub-systems dominated by intermittent energy sources, energy prices have been negative, due to the cost of shutting down wind turbines. Future systems will be dominated by intermittent energy sources, and it is completely feasible that mitigating floods by running compressors will be a net income incident for power companies in the future.

3.2. Energy Storage and Power Production

ACUR provides an additional volume for storage of water, and at some locations, this additional volume might be very useful. The world is in dire need of reliable, large-scale and fast electrical energy storage technologies. Currently, the best technology by far to provide this is Pumped Storage Power plants (PSP) in addition to large reservoirs. The PSPs are, however, dependent on having both water available as well as a steep terrain with the possibility to construct a reservoir with high elevation. Such locations are not found everywhere, and the best locations have already been developed. The expansion of the power capacity at existing PHS sites has been performed at several locations. This is simply a matter of adding units between the existing upper and lower reservoir. This increases the power that can be produced, but emptying the upper reservoir will then go faster, since the energy storage capacity has not been increased. At some point, the expansion of power at the PSP might be limited by the rate of water level decrease, because a too high rate might introduce too much sediment into the water, or landslides might occur. Installing and operating ACUR to reduce the rate of water level decrease by displacing water into the main reservoir might provide the possibility to add more power by working around this limitation. A suggested position of ACUR in conjunction to the reservoir can be seen in Figure 6.

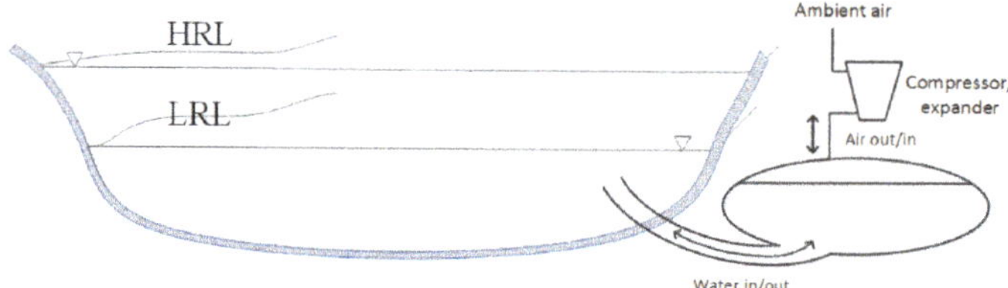

Figure 6. ACUR connected to a reservoir. HRL: highest reservoir level; LRL: lowest reservoir level.

Installing and operating ACUR thus gives the possibility to install more capacity than otherwise possible. Furthermore, the added volume of ACUR represents a novel way of increasing the energy storage capacity at reservoirs. The conventional way of doing this is not easy, because this will in most cases imply that the height of the dam must be increased, and this is not possible in most cases because the additional forces on the construction void the structural integrity of the existing dam (in Europe, this is still possible at several locations due to the high safety factor used when designing the dams due to the cold war). One project in which the dam height was increased due to energy storage increase considerations is at Vianden PHS [16]; however, this dam is not representative of the majority of dams.

ACUR represents the possibility of increasing the energy storage at an existing power plant without the need to replace the existing dam.

The position of ACUR below the lowest reservoir level (LRL) in Figure 6 is intentional, and the interested reader can find reasons for this in a previous paper by one of the authors [17]. In that same paper, the ACUR configuration is theoretically demonstrated, showing that the reduction in the efficiency of a pump–turbine cycle due to losses in the compressed air system is small, and that the relative reduction decreases with the increasing head of the PHS.

Water bodies that are currently off-limits for use as reservoirs for pumped storage power plants due to water level limitations might also be allowed for this kind of operation. Imagine two water bodies close to each other with an elevation difference that makes them suitable for pumped hydro; constructing a pumped hydro plant operating between the two water bodies and water displacement capabilities at both bodies would make it possible to operate the plant without any water level changes to either water body by the opposite operation of the displacement capabilities at the two reservoirs. As water is pumped from the lower body, displacement by air would make the level constant. At the upper water body, air would have to be evacuated to make place for the pumped water. Actually, the air could be a closed cycle as well, in which air from the upper part is moved to the lower part simultaneously as water is pumped, counter-cycled to the water. This can be seen in Figure 6, where inflatable balloons are used. ACUR could also be used, but for both applications, the storage is limited to the volume of ACUR/balloons. This might very well be suited for intra-day and hydraulic short-circuit operation for grid stability issues close to a large concentration of energy consumers. In fact, it could be possible to make existing Run-of-River (ROR) power plants operate in a less ROR-like manner. With little or no power production in the power plant, the water level effect of the inflow of water in the upper reservoir could be compensated for by removing the air from the balloons in this reservoir. The same air could be added to the balloons downstream of the powerplant, displacing water and maintaining the flow to the downstream river sections, as seen in Figure 7. When power is needed, the power plant can be ramped up or brought on-line, producing large amounts of power as the air is cycled back again, mitigating any water level and the discharge fluctuations of the reservoir and river. This would change the merit order of the power plant and provide a much-needed addition to the flexibility of the electrical system, allowing the higher penetration of intermittent renewable energy sources into the system.

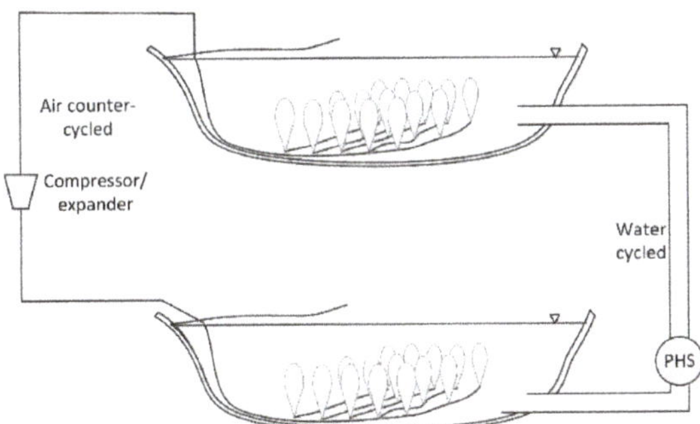

Figure 7. Schematics of counter-cycled water and air Pumped Hydro Storage (PHS).

For hydropower plants (without pumping capabilities), there might be limitations on their operation due to restrictions on the allowed water level in the reservoirs. This is the case for several Nordic hydropower plants, in which the restrictions are typically linked to the summer water level due to fish migration, recreation, transport, etc. The power production capabilities are thus greatly reduced because the level is linked to the volume, and the production of power reduces the volume and thus reduce the level. In such cases, using compressed air to displace volume would lift or ease the restriction, and the installed capacity would be available for a duration corresponding to the available maximum volume of air. Interestingly, this volume of air is then also available as a volume for flood damping, as the overflow of water can be avoided as the air is expanded, storing the flood. Flood reduction by hydropower reservoirs has great potential socio-economic value; in Norway, as one example, conservative estimates on some waterways estimate several hundreds of million Euros could be saved by the flood dampening capabilities of such reservoirs [18]. As precipitation is expected to be more intense, the capabilities for flood damping by existing hydropower reservoirs might not be sufficient. Furthermore, the Norwegian Water Resources and Energy Directorate has stated that, in a phase where concessions are being renewed, environmental and recreational concerns will be given more weight and will lead to less flexibility due to stronger water level restrictions. This has made Statkraft—the biggest producer of renewable energy in Europe—raise concerns about the future capability of flood protection in Norway [19].

Another possible use of ACUR is at storage hydropower plants with outlets to rivers. Storage power plants are very important for the balancing of the grid because they have the energy storage and power capacity needed to balance baseload generation, intermittent generation and consumption via market mechanisms. This is typically referred to as "hydropeaking" [20]. However, outlets to rivers are often problematic because the discharge to the river can violently fluctuate according to the hydropeaking of the power plant. This is a serious environmental problem because it significantly alters the natural dynamic characteristics of the river flow, affecting the ecology in a negative and often unacceptable way [21]. There are ways of reducing the effect of hydropeaking, and retention/compensation basins are examples of constructional measures [22]. However, there are limitations concerning the flexibility enabled by such basins as well, and additional measures are likely to increase the flexibility further. Using ACUR as a temporary water storage volume and compressed air as an active measure to control the flow out of ACUR will make it possible to smoothen the discharge transients at large operational changes, starts and stops. The layout of this can be seen in Figure 8.

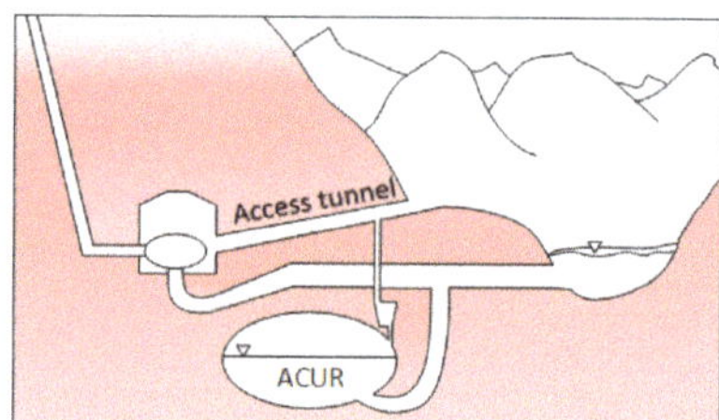

Figure 8. ACUR used to smoothen discharge to river, not to scale.

This application of ACUR is currently being investigated in the EU Horizon 2020 project HydroFlex, and the findings are that ACUR is able to smoothen the discharge well, providing the possibility of much faster ramp up/start up and ramp down/shut down of the power plants within current environmental restrictions. This can be seen in Figures 9 and 10, respectively. These results are obtained from a simulation of the power plant Bratsberg [23], which has two identical units. The "Q setpoint" is the target for the governor, which is implemented to control the compressor providing an active use of the ACUR element and represents current limitations due to environmental restrictions. "Q at discharge" is the discharge into the downstream river that results from the operation of the ACUR element. As can be seen, the start-up and shut-down of the units can be performed much faster than the targeted flow due to the presence of the governed ACUR element. The ACUR technique significantly increasing the flexibility of the units. The analysis corresponds to TRL3 and is regarded as proof of concept by the authors.

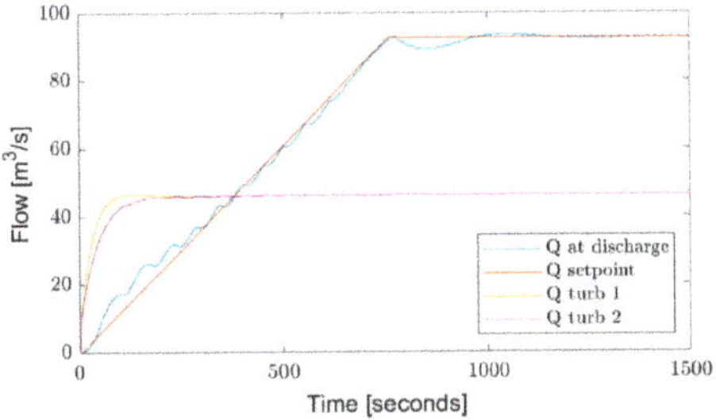

Figure 9. Smoothening of discharge by ACUR at start-up. Reprinted with permission from [23].

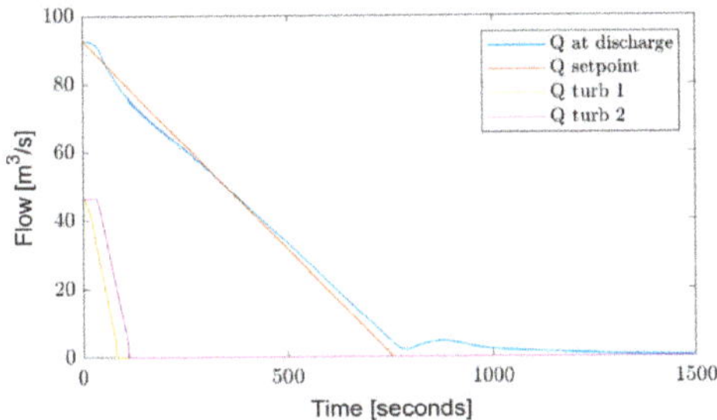

Figure 10. Smoothening of discharge by ACUR at shut-down. Reprinted with permission from [23].

3.3. Environmental Operation

Several of the applications already described might be categorized as being initiated by environmental concerns. However, other operations might give additional environmental benefits. In large hydropower and irrigation reservoirs, water can become stagnant and result in an increase in parasitic diseases [24] and other organisms associated with human illness [25]. In such cases, a grid of balloons could be used (a minimum of two if the overall water level is to be kept constant) to circulate water by inflating some balloons with a volume rate of air and at the same time deflating some balloons with the same rate. In this way, water is set in motion within the reservoir. This would contribute to preventing negative effects from occurring due to stagnant water, simply because it would be less stagnant. Such systems are not limited to rivers and power production and can be of interest for the storage of water in general. As a final remark on this, it is noted that most of the undeveloped hydropower potential is in Africa, South America and Asia [26], in climates where there are challenges associated with these kinds of problems.

4. Discussion

The authors recognize that the concepts described in this paper are far from being deployed, and much research and development is needed before the concepts reach a TRL which make it ready for field-testing. However, the authors see no major technological challenges related to these concepts. Compressors exist today that move large volumes of air against low pressures. The compression heat and subsequent energy loss is not very high for low pressures, if energy recovery from the compressed air is necessary. Constructing caverns and trapping air is understood and has been successfully applied, even at high pressures. Balloons have been made which are intended to be inflated/deflated by air, even at high pressures. Anchoring balloons to the bottom of a reservoir should be a matter of dimensioning.

5. Conclusions

Climate change is expected to make the task of future water management more difficult. The presented concepts represent a new degree of freedom, which might facilitate this task and make better use of existing infrastructure. The high degree of conceptuality presented in this paper, along with no indication of cost, makes conclusions on an applicable level difficult. However, the intention of the paper is to stimulate out-of-the-box thinking and the assessment of possible applications among engineers and researchers throughout the community. No technical challenges have been identified that makes the concepts unfeasible, but further research and development must be performed.

Author Contributions: Conceptualization, methodology, visualization and writing—original draft: P.-T.S.; funding, acquisition and project administration: T.S.L.; investigation, supervision and writing—review and editing: P.-T.S. and T.S.L.

Funding: Parts of this research is funded by European Union Horizon 2020 project HydroFlex grant agreement No 764011, namely the part on using ACUR to smoothen discharge transients from hydropower plants with outlet to rivers.

Conflicts of Interest: The authors declare no conflict of interest. The funders had no role in the design of the study; in the collection, analyses, or interpretation of data; in the writing of the manuscript, or in the decision to publish the results.

Appendix A

Referring to Figure A1, the equations leading to the power needed (Equation (7)) for the operation of the air system are developed here.

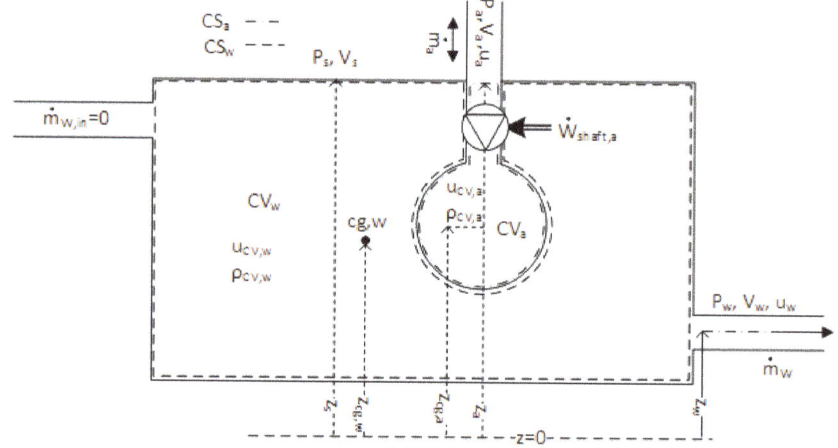

Figure A1. Schematics of the system.

Adding the two versions of Equation (6) for the systems together, we get

$$\dot{W}_{shaft,a} = \tfrac{d}{dt}\int_{CV,w}\rho e\,dV + \int_{CS,w}\rho\left(\tfrac{P}{\rho}+e\right)(\vec{V_r}\cdot\vec{n})dA + \tfrac{d}{dt}\int_{CV,a}\rho e\,dV$$
$$+ \int_{CS,a}\rho\left(\tfrac{P}{\rho}+e\right)(\vec{V_r}\cdot\vec{n})dA. \tag{A1}$$

Replacing the specific energy e with the specific energy terms it contains results in

$$\dot{W}_{shaft,a} = \tfrac{d}{dt}\int_{CV,w}\rho\left(u+\tfrac{V^2}{2}+gz\right)dV + \int_{CS,w}\rho\left(\tfrac{P}{\rho}+u+\tfrac{V^2}{2}+gz\right)(\vec{V_r}\cdot\vec{n})dA$$
$$+\tfrac{d}{dt}\int_{CV,a}\rho\left(u+\tfrac{V^2}{2}+gz\right)dV$$
$$+ \int_{CS,a}\rho\left(\tfrac{P}{\rho}+u+\tfrac{V^2}{2}+gz\right)(\vec{V_r}\cdot\vec{n})dA \tag{A2}$$

The further assumptions are as follows: fixed CS for inlets and outlets ($\vec{V_r} = \vec{V}$); all integrands are uniform and constant (allowing the integrands to be extracted from the integrals) except for z, for the water volume, which will be addressed later; the air volume expands equally in all directions, making the COG for the air volume constant, thus allowing the z for the air volume to be substituted with the average value $z_{cg,\,a}$ representing the COG; very small velocities in the volumes making the square of the velocities negligible amount to the following:

$$
\begin{aligned}
\dot{W}_{shaft,a} = {}& u_{CV,w}\rho_w \frac{d}{dt}\int_{CV,w} dV + \rho_w g \frac{d}{dt}\int_{CV,w} z\,dV \\
&+ \rho_w \left(\frac{P_w}{\rho_w} + u_w + \frac{V_w^2}{2} + g z_w\right)\int_{CS,w} dQ \\
&+ u_{CV,a}\rho_{aw}\frac{d}{dt}\int_{CV,a} dV + \rho_a g z_{cg,a}\frac{d}{dt}\int_{CV,a} dV \\
&- \rho_a \left(\frac{P_a}{\rho_a} + u_a + \frac{V_a^2}{2} + g z_a\right)\int_{CS,a} dQ
\end{aligned}
\tag{A3}
$$

All integrals of differentials end up of being the discharge of air or water, hence

$$
\begin{aligned}
\dot{W}_{shaft,a} = {}& u_{CV,w}\rho_w(-Q_{w,out}) + \rho_w g \frac{d}{dt}\int_{CV,w} z\,dV + \left(\frac{P_w}{\rho_w} + u_w + \frac{V_w^2}{2} + g z_w\right)\rho_w Q_{w,out} \\
&+ u_{CV,a}\rho_a Q_a + \rho_a g z_{cg,CV,a}Q_a - \left(\frac{P_a}{\rho_a} + u_a + \frac{V_a^2}{2} + g z_a\right)\rho_a Q_a
\end{aligned}
\tag{A4}
$$

The lack of losses that practically make the inlet/outlet internal energies equal to the internal energies inside the CVs implies that

$$
\begin{aligned}
\dot{W}_{shaft,a} = {}& \rho_w g \frac{d}{dt}\int_{CV,w} z\,dV + \left(\frac{P_w}{\rho_w} + \frac{V_w^2}{2} + g z_w\right)\rho_w Q_{w,out} + \rho_a g z_{cg,a}Q_a \\
&- \left(\frac{P_a}{\rho_a} + \frac{V_a^2}{2} + g z_a\right)\rho_a Q_a
\end{aligned}
\tag{A5}
$$

Using the Bernoulli equation [6] from the surface to the inlet and the outlet of air and water, respectively, yields

$$
\frac{P_s}{\rho_w} + g z_s = \frac{P_w}{\rho_w} + \frac{V_w^2}{2} + g z_w,
\tag{A6}
$$

$$
\frac{P_s}{\rho_a} + g z_s = \frac{P_a}{\rho_a} + \frac{V_a^2}{2} + g z_a.
\tag{A7}
$$

Substituting this back into Equation (A5) results in

$$
\dot{W}_{shaft,a} = \rho_w g \frac{d}{dt}\int_{CV,w} z\,dV + \left(\frac{P_s}{\rho_w} + g z_s\right)\rho_w Q_{w,out} + \rho_a g z_{cg,a}Q_a - \left(\frac{P_s}{\rho_a} + g z_s\right)\rho_a Q_a
\tag{A8}
$$

which may be rewritten into the following expression:

$$
\dot{W}_{shaft,a} = \rho_w g \frac{d}{dt}\int_{CV,w} z\,dV + (P_s + \rho_w g z_s)Q_{w,out} + \rho_a g z_{cg,a}Q_a - (P_s + \rho_a g z_s)Q_a.
\tag{A9}
$$

Now, the remaining integral will be addressed. This integral accounts for the rate of change of the potential energy of the water body. This energy is typically found by multiplying the mass with the COG elevation, as is done for the air system in this example. However, the emptying and displacement of water, as in this example, will potentially change the COG for the mass, as well as reducing the mass. If the displacement of water is performed above the COG level of the water, the COG will be pushed downwards. If the displacement of water is performed below the COG of the water, the COG is pushed upwards. This will thus change the potential energy of the mass and must be considered. This is done in the following.

Consider now the air volume filled with water. We can then find the potential energy of the reservoir completely filled with water by adding the integrals for the two volumes of water together. Since we know the shape of the reservoir volume, we also know the value of the COG elevation $z_{cg,res}$ for this volume V_{res}:

$$\rho_w g \int_{CV,w} z\,dV + \rho_w g \int_{CV,A} z\,dV = \rho_w g \int_{CV,res} z\,dV = \rho_w g z_{cg,res} V_{res}. \tag{A10}$$

The integral that represents the part of the volume that contains water in our case is of particular interest; thus, rearranging this, we get

$$\int_{CV,w} z\,dV = z_{cg,res} V_{res} - \int_{CV,A} z\,dV. \tag{A11}$$

Using our assumption regarding the iso-directional expansion of the air system, we can rewrite and obtain (as already done in from Equation (A3) to Equation (A4))

$$\int_{CV,w} z\,dV = z_{cg,res} V_{res} - z_{cg,\,a} \int_{CV,A} dV = z_{cg,res} V_{res} - z_{cg,A} V_A. \tag{A12}$$

We must now multiply with $\rho_w g$ and differentiate with respect to time to get the correct term to substitute into Equation (A9):

$$\rho_w g \frac{d}{dt} \int_{CV,w} z\,dV = \rho_w g \frac{d}{dt}\left(z_{cg,res} V_{res} - z_{cg,A} V_A\right) = -\rho_w g z_{cg,A} \frac{dV_A}{dt} = -\rho_w g z_{cg,A} Q_A. \tag{A13}$$

Substituting back into Equation (A9), we get

$$\dot{W}_{shaft,a} = -\rho_w g z_{cg,\,a} Q_A + (P_s + \rho_w g z_s) Q_{w,out} + \rho_a g z_{cg,A} Q_a - (P_s + \rho_a g z_s) Q_a, \tag{A14}$$

which can be arranged to

$$\dot{W}_{shaft,a} = (P_s + \rho_w g z_s) Q_{w,out} + (\rho_a - \rho_w) g z_{cg,A} Q_A - (P_s + \rho_a g z_s) Q_a. \tag{A15}$$

In our cases, we want the reservoir volume to be constant, so $Q_{w,out} = Q_a$:

$$\dot{W}_{shaft,a} = \left(\rho_w g z_s + (\rho_a - \rho_w) g z_{cg,A} - \rho_a g z_s\right) Q_{w,out}, \tag{A16}$$

which may be rewritten as

$$\dot{W}_{shaft,a} = g(\rho_w - \rho_a)\left(z_s - z_{cg,A}\right) Q_{w,out}. \tag{A17}$$

References

1. Kumar, A.; Schei, T.; Ahenkorah, A.; Rodriguez, R.C.; Devernay, J.-M.; Freitas, M.; Hall, D.; Killingtveit, Å.; Liu, Z.; Heath, G. Hydropower. In *IPCC Special Report on Renewable Energy Sources and Climate Change Mitigation*; Edenhofer, O., Pichs-Madruga, R., Sokona, Y., Seyboth, K., Matschoss, P., Kadner, S., Zwickel, T., Eickemeier, P., Hansen, G., Schlömer, S., et al., Eds.; Cambridge University Press: Cambridge, UK; New York, NY, USA, 2011; Available online: https://www.ipcc.ch/site/assets/uploads/2018/03/Chapter-5-Hydropower-1.pdf (accessed on 25 October 2019).
2. Nakićenović, N.; Alcamo, J.; Davis, G.; de Vries, B.; Fenhann, J.; Gaffin, S.; Gregory, K.; Griibler, A.; Jung, T.; Kram, T.; et al. *Special Report on Emissions Scenarios: A Special Report of Working Group III of the Intergovernmental Panel on Climate Change*; Nakićenović, N., Swart, R., Eds.; Cambridge University Press: Cambridge, UK; New York, NY, USA, 2000; ISBN 978-0-521-80081-5. Available online: https://www.ipcc.ch/site/assets/uploads/2018/03/emissions_scenarios-1.pdf (accessed on 25 October 2019).
3. Trenberth, K. Changes in precipitation with climate change. *Clim. Res.* **2011**, *47*, 123–138. [CrossRef]
4. European Commision. The Effects of Temporary State Aid Rules Adopted in the Context of the Financial and Economic Crisis. 2011. Available online: https://ec.europa.eu/competition/publications/reports/temporary_stateaid_rules_en.html (accessed on 25 October 2019).
5. International Energy Agency (IEA). World Energy Investment 2018. Available online: https://www.iea.org/wei2018/ (accessed on 25 October 2019).

6. Cengel, Y.A.; Cimbala, J.M. *Fluid Mechanics, Fundamentals and Applications*, 3rd ed.; McGraw Hill Education: New York, NY, USA, 2014; ISBN 987-1-259-01122-1.

7. Juárez, A.; Adeva-Bustos, A.; Alfredsen, K.; Dønnum, B.O. Performance of A Two-Dimensional Hydraulic Model for the Evaluation of Stranding Areas and Characterization of Rapid Fluctuations in Hydropeaking Rivers. *Water* **2019**, *11*, 201. [CrossRef]

8. Bin Asad, S.M.S.; Lundström, T.S.; Andersson, A.G.; Hellström, J.G.I.; Leonardsson, K. Wall Shear Stress Measurement on Curve Objects with PIV in Connection to Benthic Fauna in Regulated Rivers. *Water* **2019**, *11*, 650. [CrossRef]

9. De Jong, M. Commercial Grid Scaling of Energy Bags for Underwater Compressed Air Energy Storage. In Proceedings of the 2014 Offshore Energy & Storage Symposium, Windsor, ON, Canada, 10–11 July 2014. Available online: http://www.thin-red-line.com/140714_ThinRedLine_OSES2014.pdf (accessed on 25 October 2019).

10. Salvadore, E.; Bronders, J.; Batelaan, O. Hydrological modelling of urbanized catchments: A review and future directions. *J. Hydrol.* **2015**, *529*, 62–81. [CrossRef]

11. Moen, T.S. Mitigation of Discharge Fluctuations from Hydropower Plants by Active Measures. Master's Thesis, Norwegian University of Science and Technology, Trondheim, Norway, June 2018.

12. Heggenes, J.; Traaen, T. Downstream migration and critical water velocities in stream channels for fry of four salmonid species. *J. Fish Biol.* **1988**, *32*, 717–727. [CrossRef]

13. Lindmark, E.; Gustavsson, L.H. Field study of an attraction channel as entrance to fishways. *River Res. Appl.* **2008**, *24*, 564–570. [CrossRef]

14. Green, T.M.; Lindmark, E.M.; Lundström, T.S.; Gustavsson, L.H. Flow characterization of an attraction channel as entrance to fishways. *River Res. Appl.* **2011**, *27*, 1290–1297. [CrossRef]

15. McDonald, R.R.; Nelson, J.M.; Paragamian, V.; Barton, G.J. Modeling hydraulic and sediment transport processes in white sturgeon spawning habitat on the Kootenai River, Idaho. *J. Hydraul. Eng.* **2017**, *136*, 1077–1092. [CrossRef]

16. ACI Conference EU EES2 Site Visit. Available online: https://web.archive.org/web/20130206134815/http://www.wplgroup.com/aci/conferences/eu-ees2-site-visit.asp (accessed on 25 October 2019).

17. Storli, P.-T. Novel methods of increasing the storage volume at Pumped Storage Power plants. *Int. J. Fluid Mach. Syst.* **2017**, *10*, 209–217. [CrossRef]

18. Glover, B.; Sølthun, N.R.; Walløe, K.L. *Verdien av Vassdragsreguleringer for Reduksjon av Flomskader*; Multiconsult: Oslo, Norway, 2018. Available online: https://www.energinorge.no/contentassets/368e1425713a4c3a8a47fce6dd86dffe/flomrapport-22-03-2018.pdf (accessed on 25 October 2019).

19. Viseth, E.S. 70 Prosent av Norske Vannkraftverk Skal få Nye Kjøreregler—Statkraft Frykter Flomvernet Blir for Dårlig. Available online: https://www.tu.no/artikler/70-prosent-av-norske-vannkraftverk-skal-fa-nye-kjoreregler-statkraft-frykter-flomvernet-blir-for-darlig/449373 (accessed on 25 October 2019).

20. Charmasson, J.; Zinke, P. *Mitigation Measures against Hydropeaking Effects*; SINTEF: Trondheim, Norway, 2011; ISBN 978-82-594-3517-0.

21. Moog, O. Quantification of daily peak hydropower effects on aquatic fauna and management to minimize environmental impacts. *Regul. Rivers Res. Manag.* **1993**, *8*, 5–14. [CrossRef]

22. Schleiss, A.J.; Speerli, J.; Pfammatter, R. *Swiss Competences in River Engineering and Restoration*; CRC Press: Boca Raton, FL, USA, 2014; ISBN 978-1-4987-0443-4.

23. Mestvedthagen, M. Increasing Operational Flexibility of Hydropower by New Technology. Master's Thesis, Norwegian University of Science and Technology, Trondheim, Norway, June 2019.

24. Medina, D.C.; Findley, S.E.; Doumbia, S. State–Space Forecasting of *Schistosoma haematobium* Time-Series in Niono, Mali. *PLoS Negl. Trop. Dis.* **2008**, *2*, e276. [CrossRef]

25. Yirefu, F.; Tafesse, A.; Gebeyehu, T.; Tessema, T. Distribution, Impact and Management of Water Hyacinth at Wonji-Shewa Sugar Factory. *Eth. J. Weed Mgt.* **2007**, *1*, 41–52.

26. Zhou, Y.; Hejazi, M.; Smith, S.; Edmonds, J.; Li, H.; Clarke, L.; Calvin, K.; Thomson, A. A comprehensive view of global potential for hydro-generated electricity. *Energy Environ. Sci.* **2015**, *8*, 2622–2633. [CrossRef]

© 2019 by the authors. Licensee MDPI, Basel, Switzerland. This article is an open access article distributed under the terms and conditions of the Creative Commons Attribution (CC BY) license (http://creativecommons.org/licenses/by/4.0/).

Article

Smart Water Management towards Future Water Sustainable Networks

Helena M. Ramos [1],*, Aonghus McNabola [2], P. Amparo López-Jiménez [3] and Modesto Pérez-Sánchez [3]

[1] Civil Engineering, Architecture and Georesources Departament, CERIS, Instituto Superior Técnico, Universidade de Lisboa, 1049-001 Lisboa, Portugal
[2] Department of Civil, Structural and Environmental Engineering, Trinity College Dublin, D02 PN40 Dublin, Ireland; amcnabol@tcd.ie
[3] Hydraulic and Environmental Engineering Department, Universitat Politècnica de València, 46022 Valencia, Spain; palopez@upv.es (P.A.L.-J.); mopesan1@upv.es (M.P.-S.)
* Correspondence: hramos.ist@gmail.com or helena.ramos@tecnico.ulisboa.pt

Received: 2 November 2019; Accepted: 19 December 2019; Published: 21 December 2019

Abstract: Water management towards smart cities is an issue increasingly appreciated under financial and environmental sustainability focus in any water sector. The main objective of this research is to disclose the technological breakthroughs associated with water and energy use. A methodology is proposed and applied in a case study to analyze the benefits to develop smart water grids, showing the advantages offered by the development of control measures. The case study showed the positive results, particularly savings of 57 GWh and 100 Mm^3 in a period of twelve years when different measures from the common ones were developed for the monitoring and control of water losses in smart water management. These savings contributed to reducing the CO_2 emissions to 47,385 t $CO_{2\text{-eq}}$. Finally, in order to evaluate the financial effort and savings obtained in this reference systems (RS) network, the investment required in the monitoring and water losses control in a correlation model case (CMC) was estimated, and, as a consequence, the losses level presented a significant reduction towards sustainable values in the next nine years. Since the pressure control is one of the main issues for the reduction of leakage, an estimation of energy production for Portugal is also presented.

Keywords: smart water management; smart water grids; water drinking network; water losses; energy production

---

1. Introduction

1.1. Overview of the Water Sector

The water industry is subject to new challenges regarding the sustainable management of urban water systems. There are many external factors, including impacts of climate change, drought, and population growth in urban centres, which lead to an increase of the responsibility in order to adopt more sustainable management of the water sector [1]. The coverage of the costs, the monitoring of the non-revenue water (NRW) and the knowledge of the customers' demand for the fairness in revenues are some of the main challenges the water management has to solve [2]. Due to the population growth increasing and a concentration of water needs, a consequent requisite of water management is necessary. Under this reality, the use of advanced technologies, as well as the adoption of more robust management models, are necessary to better suit the water demands [3].

Over the past decades, many parts of the world witnessed the growing water demand, the risks of pollution water supply, as well as the severe water stress. The leading and irreplaceable role that water plays in sustainable development has become increasingly recognized; the management of water resources and the provision of services related to water continues to be minor in the scale of public perception and government priorities of several countries [4]. This lack of water resources is currently satisfied by the water transfer between basins, the desalinization, the regeneration of waste water, and the exploration of wells [5]. The implementation of more efficiency, the water and energy nexus, as well as the water loss control by the best pressure management and smart device implementation, would conduct a sustainable water sector.

1.2. Smart Water Management

Smart water management aims at the exploitation of water, at the regional or city level, on the basis of sustainability and self-sufficiency. This exploitation is carried out through the use of innovative technologies, such as information and control technologies and monitoring [6]. Hence, water management contributes to leakage reduction, water quality assurance, improved customer experience, and operational optimization, amongst other key performance benefits [7,8]. A smart city can be defined as the city in which an investment in human and social capital is performed, by encouraging the use of "Information and Communication Technology" (ICT) as an enabler of sustainable economic growth, providing improvements in the quality of life of consumers, and consequently, allowing better management of water resources and energy [9]. It is important to recognize that the concept of a smart city is not limited only to technological advances, but aims to promote socioeconomic development [10,11]. Through this model, a city can examine its current state and, in turn, identify the areas that require further development in order to meet the necessary conditions for a smart city [12].

The development of smart techniques requires technology use in the water systems as well as its implementations. Smart water systems are used to improve the situation of many networks characterized by degraded infrastructure, irregular supplies, and low levels of customer satisfaction or substantial deviations of the proportional bills to real consumption. A smart water system can lead to more sustainable water services, reducing financial losses, enabling innovative business models to serve the urban and rural population better [13].

Some of the main advantages of smart water management are a better understanding of the water system, detection of leaks, conservation, and monitoring of water quality. The implementation of smart water system technologies enables public services companies to build a complete database for the identification of the areas where water losses or illegal connections occur. The advantages of smart water grids are economic benefits to water and energy conservation, while the efficiency of the system can improve customer service. The wireless data transmission allows the customers to analyze their water consumption towards preserving and reducing the water bill, in some cases above 30% [14]. Some of the main technologies are be listed as follows:

(i) Smart pipe and sensor; a smart pipe is designed as a module unit with a monitoring capacity expandable for future available sensors [15]. With several smart pipes installed in critical sections of a public water system, real-time monitoring automatically detects the flow, the pressure, leaks and water quality, without changing the operating conditions of the hydraulic circuit. Briefly, the smart wireless sensor network is a viable solution for monitoring the state of pressure and loss of water control in the system. The main advantage compared to other methods of water loss control is the continuous monitoring of the network without local operator intervention and with low energy consumption of the wireless sensor, allowing to remain operational for long periods [16].

(ii) Smart water metering; a smart meter is a measuring device that can store and transmit the consumption with a certain frequency. To develop an efficient water management system, it is necessary to install sensors and/or actuators to monitor the water systems [17]. Therefore, while water meters can be read monthly or one reading every two months with the water bill generated from this manual reading, the smart metering can obtain the consumption at long distance and with a high frequency, providing instant access to the information for customers and managing entities. The management of this information requires an advanced metering infrastructure (AMI), and therefore, the water companies should install this in order to improve hydraulic and energy efficiency, enabling leakage control as well as illegal connections in terms of water volumes [18].

(iii) Geographic Information System (GIS); GIS plays a strong role in smart water management, providing a complete list of the components along the network and their spatial locations. GIS becomes essential for the management of the water systems, allowing the inclusion of the spatial components in an oriented model to improve the planning and management through a clear evolution of spatial constituents in the network. The major advantage of GIS is the simulation of reality based on data systems designed to collect, store, receive, share, manipulate, analyze, and present information that is geographically referenced [19,20].

(iv) Cloud computing and supervisory control and data acquisition (SCADA); it is referred to the use of memory and storage capacities and calculation of computers and servers shared and linked through the internet, by following the code of network computing. Cloud computing is defined as "a new style of computing in which the resources are dynamically scalable and often virtualized being provided as a service over the internet" [21], such as large repositories of virtualized resources, hardware, development platforms, and software, with easy access and dynamically configured to adapt to different workloads in order to optimize their use. In general, the majority of public water services make supervision, control, and data management through a SCADA system [22,23].

(v) Models, tools of optimization, and decision support systems; the implementation of a common framework for measuring the performance based on a set of relevant indicators and data applications and interfaces to support the decision of the managing entities allows the interested parties to evaluate, create trust and confidence, and monitor the improvements [24,25]. The knowledge of reliable short-term demand forecasting patterns is crucial to develop approach models and, therefore, positive decisions in real time to be implemented in smart water systems [26]. These models are focused on simulations, such as Epanet [27] or WaterGems [28]. These tools can be supported using optimization techniques. The programming models can use simulated annealing techniques [29], fuzzy linear programming [30], and multi-objective genetic algorithms in real time [31], among others.

The objective of this research is to disclose the technological breakthroughs associated with water use and the innovations according to the monitoring of water and energy losses, proposing a strategy to improve the efficiency of the system in economic and sustainable terms. The methodology is applied to a real case study of water distribution system (named reference system (RS) due to confidential restrictions). In this network, the water company implemented measures for the monitoring of several parameters, including the water loss control associated with smart water management. This case study was compared using a correlation model case (CMC) in order to predict the benefits of similar actions used in the RS in the CMC, which was proposed in this research using a real database.

2. Materials and Methods

2.1. Brief Description of Case Study

Currently, there are technological solutions capable of supporting the management of smart water systems with a high level of efficiency, associated with the reduction of water losses and, consequently, operational costs.

On the one hand, RS integrates several subsystems of water sources, pumping stations, and treatment plants. The RS is composed of approximately 1400 km of pipes, with more than 100,000 service connections, 14 reservoirs, and 10 pumping stations, which allows storage of more than 400,000 m^3. The network is modelled by GIS in which the maintenance service is developed according to leakage and break occurrence, interrelating with the customer management database. The network is divided into four different zones that depend on topographic levels. These zones are: low level, which supplies between 0 and 30 m; medium level, which is determined between 30 and 60 m; high level, when the altitude varies between 60 and 90 m; and upper level, which supplies the area above 90 m. It is mandatory that RS has to guarantee a consumed volume of 192 Mm^3. However, the NRW was not satisfactory, and the volume stabilized around 50 Mm^3. This volume shows the high volume of losses in the distribution network. The data collection in different years over time was provided by the water company.

Due to heavy losses in the RS distribution system noticeable during the night period, RS set the ambitious goal of reducing the NRW to sustainable values, with the mark of water losses set at less than 15% by 2009 (Figure 1).

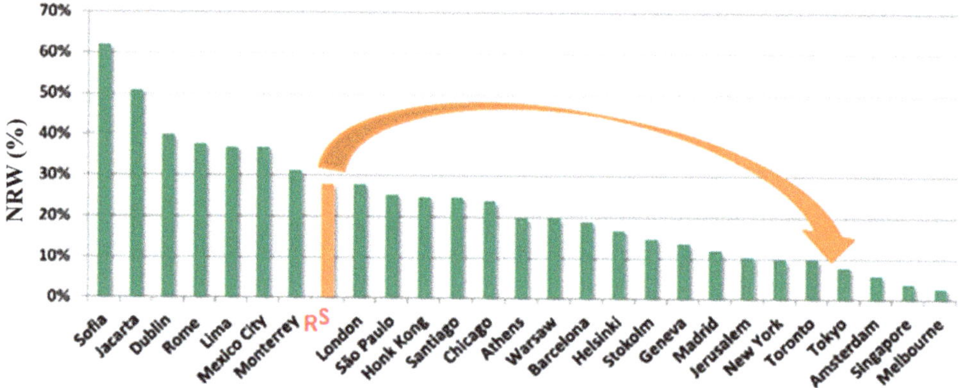

Figure 1. More efficient cities in terms of non-revenue water in the 1990s (based on [11]).

In order to reduce the water losses in 12 years for values less than 15%, RS adopted a well-defined strategy that was focused on: (i) segmentation and continuous monitoring of the network; (ii) development of analysis using internal resources; (iii) optimization of the process of active water losses control; (iv) continuous improvement based on the experience and results; (v) definition of what really is primordial in real cost (investments) control. The reduction of NRW was carried out on both leakage and illegal connections, which were detected through intensive monitoring and metering of the water distribution network.

On the other hand, CMC corresponds to a water distribution in another municipality. The system supplies about 152,000 customers. The water distribution system is composed of 6 reservoirs, which correspond to a total storage capacity of 125,450 m^3. The network has 760 km in pipes. The distribution network has approximately 64,000 service connections. The supply is almost entirely gravitational, only actively maintaining the pumping station to fill areas of higher level. At the moment, the distribution network of that city is divided into 18 DMA. The company has opted for the partition of the distribution network through the creation of interior sub-DMA, to be possible to carry out more effective monitoring and consumption control. Then, DMAs are subdivided into 31 sub-DMA.

2.2. Parameters Definition in RS

The volume of water in the RS, whether imported or extracted, is divided into billed water (BW) and NRW and even between controlled and uncontrolled consumption [32]. The billed water is the consumed water that is directly charged to customers. The NRW is the volume that includes the water losses and the consumed volume by the authorized agents (e.g., social services, fire-fighting services). A simplifying schematic of this water balance is shown in Figure 2.

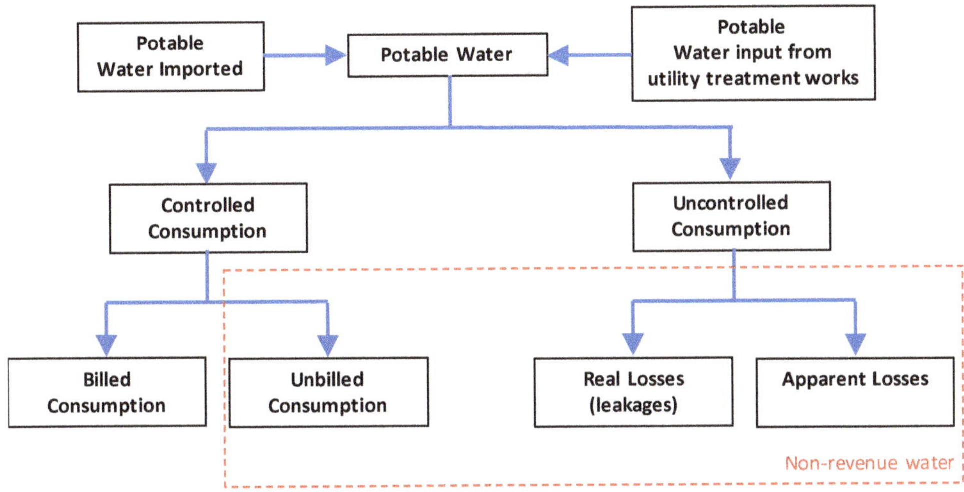

Figure 2. Distribution of the water balance in a drinking system.

Water losses reflect a measure of the quality of management and operation of each system and, consequently, of any water company. Figure 2 shows there are two types of water losses: apparent and real. The apparent or economic losses correspond to the illegal consumption, while the real losses correspond to water losses, ruptures or burst pipelines, reservoirs or service connections up to the point where the customer is connected, and water evaporation in reservoirs. The apparent and real loss volume cannot be exactly separated, and therefore, the improvement of their values should be made using recorded readings as well as increasing the monitoring (i.e., water metering) of the water network. Regarding the water losses control, this proposed strategy shows how to reach the economic level of leakage (ELL) [33]. ELL constitutes the objective value of water companies, in an attempt to minimize the overall costs associated with the water loss. This strategy means the search for the maximum investment, which is feasible compared to the cost of lost water.

The analyzed case studies (i.e., RS and CMC) propose the reduction of water losses in the water distribution system in order to improve the NRW. RS had to improve the monitoring and control of water losses. Thus, RS developed key tools for the deployment of a monitoring system, keeping the system in good operation, with respect to quantity and quality. RS considered the following technologies: (i) Geographical Information System (GIS); (ii) Management Information System for Customers (MISC); (iii) Digital Terrain Model (DTM); (iv) District Metering Areas (DMA); and (v) Hydraulic System Model (HSM).

The application of this strategy made it possible to obtain advantages in terms of quantity and quality system efficiency based on the information available regarding the system operation. In addition, it enabled the identification of consumptions of each DMA, abnormal night consumption, and the management and control of pressure in the water distribution network.

The International Water Association (IWA) recommends that a DMA should have between 1000 and 3000 customers, but in urban areas with high population density, a DMA may group together more than 3000 customers, with a maximum limit of 5000 customers [33]. Hence, the case study was subdivided according to the size of DMA into three different categories: small (DMA with less than 1000 customers), medium (between 1000 and 3000 customers), and large (more 3000 customers). These values have been tested and validated. In RS, the distribution network is divided into 150 different DMA*s*.

In order to carry out the collecting, management, and processing of the information of the water supply system, several registration devices and emitters of data, in particular, data-logging equipment and modems, were used. These devices enabled the automatic collecting data of water consumption, pressure variation, and flow or quality probes directly installed on the network. The data collected are transmitted remotely, through such devices to a central database, where they are stored, offering to the managing entity scans and frequent and reliable records, reducing the need for estimations.

The main objective of an optimization network efficiency (ONE) is to support the strategy focusing on the efficiency and the reduction of losses, providing performance indicators of each DMA. Thus, ONE integrates the process of optimization and improvement of the efficiency of a distribution network including: (i) metering and telemetry; (ii) leakage level definition; and (iii) leakage detection and repair. The implementation of flow meters at the entrance and the exit of each DMA, and the methodologies of innovative strategies for the detection and location of leaks were essential for rapid action and consequent reduction of water and energy losses in the system.

2.3. Methodology to Develop CMC

The investment analysis in monitoring and water loss control is extrapolated to another distribution system (CMC). Figure 3 shows the proposed methodology to develop the analysis.

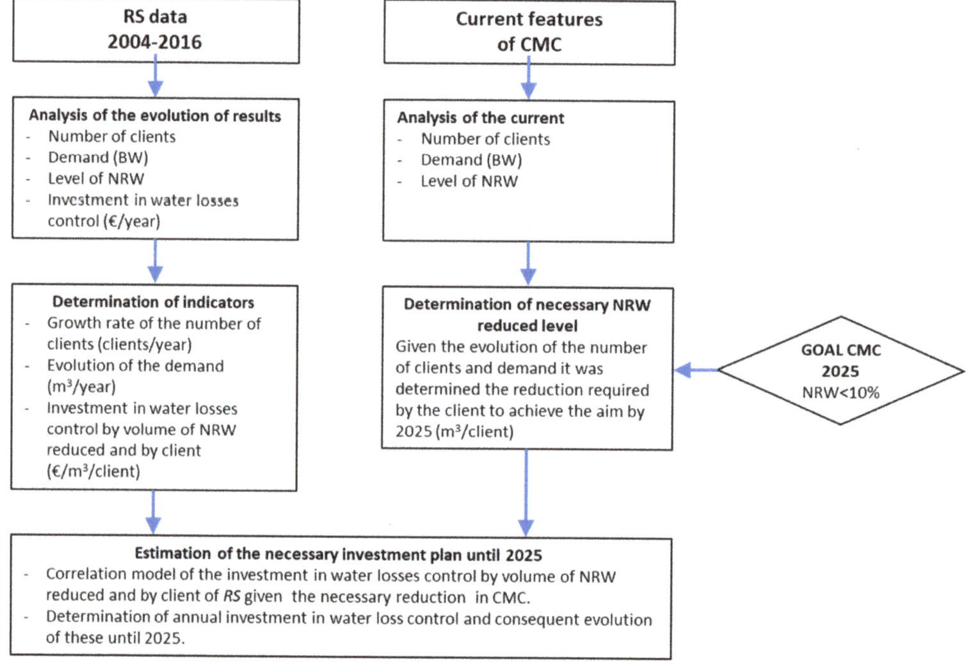

Figure 3. Flowchart of the methodology to apply the correlation model from reference systems (RS) to correlation model case (CMC).

Hence, the correlation model case (CMC) took the following procedure in the developed study:

- Analysis of the evolution of the consumption results: this analysis should be done in both systems (RS and CMC). On the one hand, the study related to RS should be developed. To do so, the knowledge of the number of customers, the BW, the current level of NRW, as well as the investment in water losses control, is necessary to know. This information is given by the water company. On the other hand, for CMC, the same information is also necessary.

- Determination of indicators: the second stage of the methodology consists of an analysis of the performance indicators, which will be used as variables in the CMC. For RS, the selected indicators were: growth rate of the number of customers per year, evolution of the water demand by the users; investment of the water company to reduce the water losses and to control the NRW. In contrast, the CMC should establish the deadline to reach the aim, as well as the limit of NRW. In the proposed case study, the deadline was 2025, and the objective value for NRW was 10%.

- Estimation of the necessary investment plan until the deadline: defined the consumption results of the water systems (input data) as well as the determination of the indicators and the defined objective for NRW, a correlation model, and the determination of the annual investment is necessary to develop. The investment plan is focused on the water losses control as well as reducing the unbilled water.

Regression analysis involves the identifying of the relationship between a dependent variable and one or more independent variables. A model of the relationship is hypothesized and estimates the parameter values that are used to develop an estimated regression equation. A simple linear regression is defined by equation (1):

$$\hat{Y} = a + b_1 X_1 + \ldots + b_p X_p \tag{1}$$

but it differs as to whether the X variables are considered random or fixed [34], and $a, b_1, \ldots, b_p$ are coefficients of the correlation equation, 'p' being the number of variables. In this statistical correlation model, random values of the X variables are considered. These variables are obtained in the sample, while the number of cases obtained at each level of the X variables is random. Hence, another sample from the same population would yield a different set of values of X and different probability distributions of X. In the (fixed) regression model, the values of X and their distributions are assumed to be, in the sample, identical to that in the population. Some experts have argued that the correlation coefficient is meaningless in a regression analysis since it depends, in large part, on the fixed particular values of X obtained in the sample and the probability distribution of X [34]. While this relationship between r and the distribution of X in the sample is certainly true, it does not necessarily follow that R and R^2 are not useful statistics in a regression analysis.

The fixed regression model fits best with experimental research where there are arbitrarily particular values of the X variables and particular numbers of cases at each value of each X. Hence, the fixed regression model is most often called the analysis of variance model. However, it is true that it is common practice to apply the regression model to data where the X variables are clearly not fixed.

When time or frequency is used to get a confidence interval, i means the joint distribution of X and the Y is bivariate (or multivariate) normal. When the distribution is bivariate normal, then it is also true that the marginal distributions of X and Y are both normal.

In the optimization of a water distribution network through the improvement of the monitoring and control of losses, the investment required in order to be possible to obtain an equivalent level of performance in other water systems needs to be estimated using a correlation model type.

A statistical analysis using the key indicators of the RS results was developed for the CMC in order to determine the annual growth rates of the number of customers and billed water, and the investment in water losses control per reduced volume of NRW and per client. Note that the determination of those parameters per client is essential to correlate different sizes of water companies.

The correlation between the annual investments per client with the decrease of NRW by the client of the following year is analyzed. This analysis would be able to determine a logistical regression for the investment required per client to achieve a certain level of NRW. Hence, it was determined that the parameter's investment in the water losses control per volume of NRW reduced and per client with the average value of annual investment by the reduction of NRW of the following year and per client by the equation (2):

$$\overline{\text{INV}}_{\text{NRW}} = \frac{\sum_{i=1}^{n} \frac{\text{INV}_i}{\text{NRW}_{i+1}}}{n} \tag{2}$$

where $\overline{\text{INV}}_{\text{NRW}}$ is the annual investment average on water losses control by a decrease of NRW and by client; INV_i is the investment on the water losses by client in the year i; NRW_{i+1} is the non-revenue water per client of the year $i + 1$.

Finally, when the correlation was developed, the economic and energy savings were determined according to water metering. Furthermore, the reduction of NRW can contribute to improving other sustainability indicators, such as social, environmental, energy, economy, and technical indicators [35]. Regarding the energy indicators, the improvement affects the network energy efficiency (IEE), excess of supplied energy (ISE), energy dissipation (IED), annual consumed energy (IAE), consumed energy per unit volume (IEFW), energy cost per unit volume implemented (IEC), energy efficiency of pumps when the system is a pump system (IEB), among others. Also, other indicators related to the consumed water will be reduced if the NRW is reduced. A significant indicator associated with the environment is $CO_{2\text{-eq}}$ since there is a direct relation between the energy (kWh) consumed in the network and the CO_2 emissions (between 582 and 877 gCO_2/kWh) [36]. Therefore, the reduction of NRW has high significance in the environmental impact of the water distribution cycle.

The proposed methodology can be used in other case studies when the water managers have enough information available (e.g., recorded and stored data, monitoring of the water systems) to analyze and to implement the corrective measures towards a more efficient water system.

3. Results

The analysis of the RS case study was carried out based on the results considering a suitable time interval as the most relevant for assessing the effects of the implemented measures in the distribution system. The implementation of the monitoring measures and the active water loss control, allowed RS to reduce the losses in the system from 20% in 2004 to less than 10% of the total volume captured in 2014 (Figure 4).

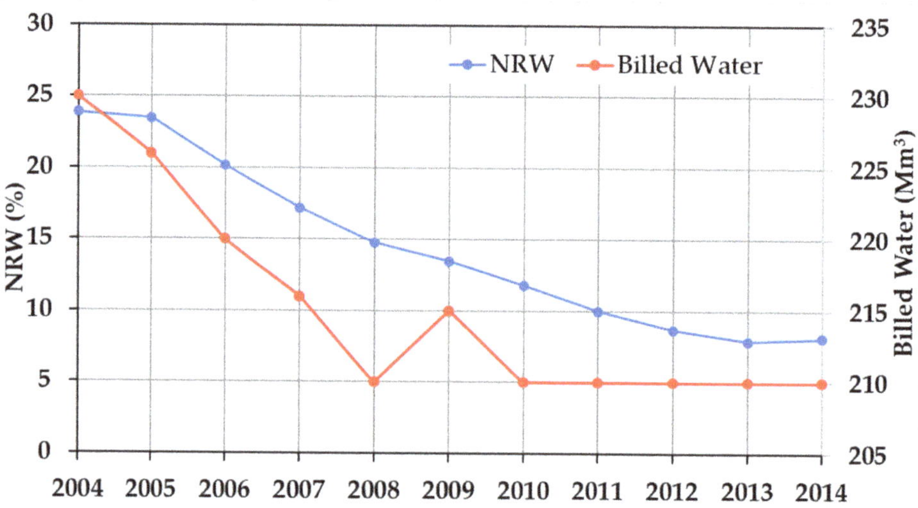

Figure 4. Non-revenue water (NRW) and billed water (BW) evolution at RS.

This decrease of NRW over 12 years was mainly due to the control of the losses in the distribution system (low level zone). In contrast, the BW decreased more than 20 Mm^3 in this period in the distribution system due to environmental concern of society, the leakages control, as well as the reduction of the unbilled water. The policy of monitoring and water loss control of RS was focused in particular on the distribution system (low level zone) since the NRW level was too high in comparison to the treatment and transport of water. The considered goal was ambitious, reducing the NRW in the distribution network for sustainable values, setting a goal of water losses less than 9% by

2016. This work positioned RS in the fifth position of the more efficient cities worldwide (Figure 5). The reduction of water losses was 67.85% compared to values of 1990 (Figure 1), which were considered as the best reference for this water company.

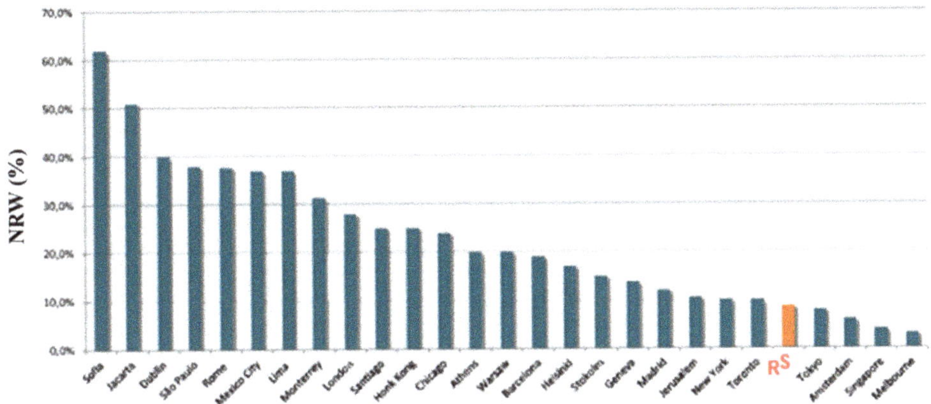

Figure 5. More efficient Cities at NRW level in 2016 (based on [11]).

Furthermore, RS still had a decrease in operating costs to the supply network. Despite the reduction of these costs, the unit cost of water produced was not sensitive to this variation and remained close to 0.30 €/m^3. This is mainly due to fixed costs of the water supply network, the decrease in demand, and an increase of unit costs of external supplies and services (ESF) in these years, in particular, the electricity.

Still, the energy bill, which is the main constituent of the ESF, contradicted the trend of growth in the market, due to the associated gains with the energy optimization enabled by monitoring and water loss control. In 12 years, RS obtained an energy saving of approximately 65 GWh, reducing the energy bill from more than €6.5 million. In addition to the energy reduction, another more direct result and representative of this policy of monitoring and water losses control was the reduction of the levels of NRW in the network, which allowed a saving close to 200 Mm^3 (€60 million) in 12 years.

These results demonstrate the improvement of the efficiency of the RS water network, in which €66 million was saved in 12 years. To reach this saving, the investment in monitoring (e.g., smart watering devices, pressure sensors, communication devices), as well as the control of water losses, were necessary. This investment was €20 million in 12 years. This value was around 30% of the total revenue in this period. This reduction of the consumed energy contributed with a theoretical reduction of 47,385 tCO$_{2\text{-eq}}$ according to [36]. The development of this strategy enabled to reduce the overall costs in the operation of the network further, offering a saving of about €46 million in 12 years of operation.

The RS data was used to estimate the major socio-economic indicators to determine the investment required to reach a certain NRW level. The annual growth rate of the number of customers was obtained through the average annual growth recorded by RS from 2004 to 2014, using a growth rate of 0.3% for the developed correlation model. In order to determine the BW progression, firstly, a canonical correlation was prepared considering the evolution of the number of customers in the distribution system in an attempt to assess the dependence of billed water with the number of customers. The obtained regression models are shown in Table 1.

Table 1. R^2 obtained in the different regression models developed.

Regression Type	Increase of Customer Number	Increase of Bill Water	Unbilled Water Investment per Customer
Linear	0.25	0.01	0.03
Logaritmich	0.33	0.02	0.14
Second Degree Polynomial	0.46	0.24	0.23
Third Degree Polynomial	0.61	45	0.34

After the determination of the correlation model indicators (Table 2), it was possible to relate the annual decrease of NRW with the annual investment in the water losses control required in the previous year. This correlation enabled an investment plan and the evolution of the distribution network features for the next nine years.

Table 2. Assessment of characteristic parameters in CMC, based on RS.

Main Features		CMC		RS	
		2016	2025	2004	2014
Total Annual Volume	(Mm^3)	20.82	17.3	127.0	101.12
BW	(Mm^3)	16.94	15.5	96.6	92.94
NRW	(Mm^3)	3.87	1.7	30.4	8.18
	(%)	18.6	10.0	23.9	8.1
Total Customers	-	150,812	155,293	339,111	349,151

The total investment required in the CMC, considering the indicators previously determined, was approximately €9.5 million for that period, allowing a reduction of more than 2.6 Mm^3 of NRW in this period. Table 2 shows the main values obtained from the correlation model.

The obtained investment parameter on the water losses control was 3.6 €/m^3 per client and year. In CMC, it was even necessary to determine the volume corresponding to the NRW goal level in 2025. From the BW and the NRW level planned, it is possible to determine the volume of NRW and water in the system for the year 2025. It was determined that these variables are independent, making, from the outset, the demand for a multivariate model in terms of the number of customers and BW. The search model has allowed for determining the evolution of the number of customers and the volume of BW at the CMC system.

4. Conclusions

In the last few years, the water sector has faced significant challenges, in particular, the effort to develop a smart water system in order to improve efficiency and sustainability performance (e.g., social, technical, and environmental). The developed designations, as well as the analyzed case studies, show that the application of this smart technology does not only contribute to the future of smart cities in terms of water but also to energy nexus, through adequate smart water planning and management.

This application will improve the water sustainability and management, as well as the policy of smart cities adequately adapted considering different constrains. The selected techniques and actions depend on the considered threshold, the capital investment, and the availability of techniques and equipment. In addition, these applied strategies must be associated with a worldwide awareness of society to the sustainable planning and management for the best use of available resources. Through the technological innovations, the smart cities will reduce costs, increase the service quality and optimize the operation of the system. The proposed methodology can also be applied to other water networks contributing to improving system efficiency and sustainability by better management of the water resources.

This research analyzed a real water system named RS (due to the confidentiality of data). In this case study, the results achieved show the implementation of measures for the monitoring and water losses control, which allowed accessing a high level of efficiency, especially the reduction of water losses and the consequent reduction of associated costs. The application of these strategies enabled changing the category of the most efficient cities level, varying its worldwide position from 20th to 5th. However, the calculation of the ELL is sensible for network changes, regional legislation, type of consumption, repair costs, ESF, and the macroeconomic situation of each country. Therefore, although these variables change, the payback period of the investment and the development of strategies to reduce the water losses are viable.

Finally, regarding the excellent results obtained in RS, the necessary investment was estimated to achieve the goal of water losses of 10%, by 2025. So, the CMC was developed and applied, presenting initially with a high level of losses of around 21.5%, requiring a total investment of around €9.5 million until 2025. In some regions, the high level of losses and the need of pressure control also allow the development of complementary solutions based on the implementation of micro-hydropower solutions using by-passes to existing PRVs or at inlets or outlets of tanks and reservoirs. This evaluation demonstrates that a significant potential for technically and economically viable micro-hydropower installations exists, which could make valuable contributions to the energy efficiency and CO_2 emission gains in the water sector.

Author Contributions: The author H.M.R. contributed with the idea and to the revision of the document and supervised the whole research. M.P.-S. contributed to the correlation model and the analysis of case studies. P.A.L.-J. and A.M. were involved in the revision and suggested guides towards the developed analyses. All authors have read and agreed to the published version of the manuscript.

Funding: This research received no external funding.

Acknowledgments: The authors wish to thank the project REDAWN (Reducing Energy Dependency in Atlantic Area Water Networks) EAPA_198/2016 from INTERREG ATLANTIC AREA PROGRAMME 2014-2020, CERIS, EPAL and ERSAR for the data availability. The authors also thank Teresa Zawerthal for collecting data and the MSc thesis based-study developments under the supervision of Helena M. Ramos.

Conflicts of Interest: The authors declare no conflict of interest.

References

1. Sachidananda, M.; Patrick Webb, D.; Rahimifard, S. A concept of water usage efficiency to support water reduction in manufacturing industry. *Sustainability* **2016**, *8*, 1222. [CrossRef]
2. Boyle, T.; Giurco, D.; Mukheibir, P.; Liu, A.; Moy, C.; White, S.; Stewart, R. Intelligent metering for urban water: A review. *Water* **2013**, *5*, 1052–1081. [CrossRef]
3. Baptista, J.; Pires, J.; Alves, D.; Marques, S.; Aleixo, C.; Andrade, I.; Ramos, L. *Relatório Anual do Sector de Águas e Resíduos em Portugal*; ERSAR–Entidade Reguladora dos Serviços de Águas e Resíduos: Lisbon, Portugal, 2016; Volume 1, 186p.
4. Ritzema, H.; Kirkpatrick, H.; Stibinger, J.; Heinhuis, H.; Belting, H.; Schrijver, R.; Diemont, H. Water management supporting the delivery of ecosystem services for grassland, heath and moorland. *Sustainability* **2016**, *8*, 440. [CrossRef]
5. Pérez-Sánchez, M.; Sánchez-Romero, F.J.; López-Jiménez, P.A. Nexo agua-energía: Optimización energética en sistemas de distribución. Aplicación "Postrasvase Júcar-Vinalopó", España. *Tecnología y Ciencias del Agua* **2017**, *8*, 19–36. [CrossRef]
6. Tadokoro, H.; Onishi, M.; Kageyama, K.; Kurisu, H.; Takahashi, S. Smart water management and usage systems for society and environment. *Hitachi Rev.* **2011**, *60*, 164–171.
7. Howell, S.; Rezgui, Y.; Beach, T. Integrating building and urban semantics to empower smart water solutions. *Autom. Constr.* **2017**, *81*, 434–448. [CrossRef]
8. Mounce, S.R.; Pedraza, C.; Jackson, T.; Linford, P.; Boxall, J.B. Cloud based machine learning approaches for leakage assessment and management in smart water networks. *Procedia Eng.* **2015**, *119*, 43–52. [CrossRef]
9. Lombardi, P.; Giordano, S.; Farouh, H.; Yousef, W. Modelling the smart city performance. *Innov. Eur. J. Soc. Sci. Res.* **2012**, *25*, 137–149. [CrossRef]

10. Colldahl, C.; Frey, S.; Kelemen, J. Smart Cities: Strategic Sustainable Development for an Urban World. Sweden: School of Engineering, Blekinge Institute of Technology. 2013. Available online: https://www.diva-portal.org/smash/get/diva2:832150/FULLTEXT01.pdf (accessed on 18 July 2017).

11. Giffinger, R.; Fertner, C.; Kramar, H.; Kalasek, R.; Pichler-Milanović, N.; Meijers, E. Smart Cities: Ranking of European Medium-Sized. Vienna, Austria: Centre of Regional Science (SRF), Vienna University of Technology. 2007. Available online: http://www.smart-cities.eu/download/smart_cities_final_report.pdf (accessed on 16 July 2017).

12. Nam, T.; Pardo, T.A. Conceptualizing smart city with dimensions of technology, people, and institutions. In Proceedings of the 12th Annual International Digital Government Research Conference: Digital Government Innovation in Challenging Times, New York, NY, USA, 12–15 June 2011; pp. 282–291.

13. Hellström, D.; Jeppsson, U.; Kärrman, E. A framework for systems analysis of sustainable urban water management. *Environ. Impact Assess. Rev.* **2000**, *20*, 311–321. [CrossRef]

14. Martyusheva, O. Smart Water Grid. USA: Department of Civil and Environmental Engineering, Colorado State University. 2014. Available online: http://www.engr.colostate.edu/~{}pierre/ce_old/Projects/Rising%20Stars%20Website/Martyusheva,Olga_PlanB_TechnicalReport.pdf (accessed on 1 August 2017).

15. Lin, Y.F.; Liu, C.; Whisler, J. Smart pipe—nanosensors for monitoring water quantity and quality in public water systems. In *World Environmental and Water Resources Congress 2009: Great Rivers*; Steve Starrett; ASCE: Reston, VA, USA, 2009; pp. 1–8.

16. Alliance for Water Efficiency. Smart Metering Introduction. Obtained on 12 August 2015, from Alliance for Water Efficiency. 2010. Available online: http://www.allianceforwaterefficiency.org/smart-meter-introduction.aspx (accessed on 30 July 2017).

17. Ntuli, N.; Abu-Mahfouz, A. A simple security architecture for smart water management system. *Procedia Comput. Sci.* **2016**, *83*, 1164–1169. [CrossRef]

18. Britton, T.C.; Stewart, R.A.; O'Halloran, K.R. Smart metering: Enabler for rapid and effective post meter leakage identification and water loss management. *J. Clean. Prod.* **2013**, *54*, 166–176. [CrossRef]

19. Sharvelle, S.; Dozier, A.; Arabi, M.; Reichel, B. A geospatially-enabled web tool for urban water demand forecasting and assessment of alternative urban water management strategies. *Environ. Model. Softw.* **2017**, *97*, 213–228. [CrossRef]

20. Furht, B.; Escalante, A. *Handbook of Cloud Computing*; Springer: New York, NY, USA, 2010.

21. EPA. Distribution System Water Quality Monitoring: Sensor Technology Evaluation Methodology and Results.A Guide for Sensor Manufacturers and Water Utilities. Ohio: EPA–Environmental Protection Agency. 2009. Available online: https://www.epa.gov/sites/production/files/2015-06/documents/distribution_system_water_quality_monitoring_sensor_technology_evaluation_methodology_results.pdf (accessed on 11 August 2017).

22. Boyer, S. SCADA: Supervisory Control and Data Acquision. USA: ISA–The Instrumentation, Systemas and Automation Society. 2004. Available online: https://www.fer.unizg.hr/_download/repository/SCADA-Supervisory_And_Data_Acquisition.pdf (accessed on 20 July 2017).

23. Airaksinen, M.; Pinto-Seppa, I.; Piira, K.; Ahvenniemi, H.; Huovila, A. Real-Time decision support systems for city management. In *Smart City–Research Highlights*; Airaksinen, M., Kokkala, M., Eds.; VTT Technical Research Centre of Finland Ltd.: Espoo, Finland, 2015; pp. 21–38. Available online: http://www.vtt.fi/inf/pdf/researchhighlights/2015/R12.pdf (accessed on 24 July 2017).

24. Boulos, P.; Wiley, A. Can we make water systems smarter? Opflow. 2013. Available online: http://innovyze.com/news/showcases/SmartWaterNetworks.pdf (accessed on 19 July 2017).

25. Gurung, T.R.; Stewart, R.A.; Beal, C.D.; Sharma, A.K. Smart meter enabled water end-use demand data: Platform for the enhanced infrastructure planning of contemporary urban water supply networks. *J. Clean. Prod.* **2015**, *87*, 642–654. [CrossRef]

26. Romano, M.; Kapelan, Z. Adaptive water demand forecasting for near real-time management of smart water distribution systems. *Environ. Model. Softw.* **2014**, *60*, 265–276. [CrossRef]

27. Rossman, L.A. *EPANET 2 User's Manual*; U.S. Environmental Protection Agency (EPA): Cincinnati, OH, USA, 2000.

28. Nazari, A.; Meisami, H. *2008 Instructing WaterGEMS Software Usage*; Water Online: Exton, PA, USA, 2008.

29. Samora, I.; Franca, M.; Schleiss, A.; Ramos, H. Simulated annealing in optimization of energy production in a water supply network. *Water Resour. Manag.* **2016**, *30*, 1533–1547. [CrossRef]

30. Sanchis, R.; Díaz-Madroñero, M.; López-Jiménez, P.A.; Pérez-Sánchez, M. Solution approaches for the management of the water resources in irrigation water systems with fuzzy costs. *Water* **2019**, *12*, 2432. [CrossRef]

31. Campos, J.A.; Jiménez-Bello, M.A.; Alzamora, F.M. Real-Time energy optimization of irrigation scheduling by parallel multi-objective genetic algorithms. *Agric. Water Manag.* **2020**, *227*, 105857. [CrossRef]

32. Sardinha, J.; Serranito, F.; Donnelly, A.; Marmelo, V.; Saraiva, P.; Dias, N.; Rocha, V. Controlo Ativo de Perdas de Água. Lisboa: EPAL–Empresa Portuguesa das Águas Livres. 2015. Available online: http://www.epal.pt/EPAL/docs/default-source/epal/publica%C3%A7%C3%B5es-t%C3%A9cnicas/controlo-ativo-de-perdas-de-%C3%A1gua.pdf?sfvrsn=30 (accessed on 31 July 2017).

33. Ndirangu, N.; Chege, A.; de Blois, R.J.; Mels, A. Local solutions in non-revenue water management through north-south water operator partnerships: The case of Nakuru. *Water Policy* **2013**, *15*, 137–164. [CrossRef]

34. Cohen, J.; Cohen, P. *Applied Multiple Regression/Correlation Analysis for the Behavioral Sciences*, 2nd ed.; Lawrence Erlbaum: Mahwah, NJ, USA, 2014.

35. Romero, L.; Pérez-Sánchez, M.; Amparo López-Jiménez, P. Improvement of sustainability indicators when traditional water management changes: A case study in Alicante (Spain). *AIMS Environ. Sci.* **2017**, *4*, 502–522. [CrossRef]

36. Spadaro, J.V.; Langlois, L.; Hamilton, B. Greenhouse Gas Emissions of Electricity Generation Chains: Assessing the Difference. *IAEA Bull.* **2000**, *42*, 19–28.

© 2019 by the authors. Licensee MDPI, Basel, Switzerland. This article is an open access article distributed under the terms and conditions of the Creative Commons Attribution (CC BY) license (http://creativecommons.org/licenses/by/4.0/).

Article

Inline Pumped Storage Hydropower towards Smart and Flexible Energy Recovery in Water Networks

Helena M. Ramos [1], **Avin Dadfar** [1], **Mohsen Besharat** [1,*] and **Kemi Adeyeye** [2]

[1] Department of Civil Engineering, Architecture and Georesources, CERIS, Instituto Superior Técnico, University of Lisbon, 1049-001 Lisbon, Portugal; hramos.ist@gmail.com or helena.ramos@tecnico.ulisboa.pt (H.M.R.); avin.dadfar@tecnico.ulisboa.pt (A.D.)

[2] Department of Architecture and Civil Engineering, University of Bath, Bath BA2 7AY, UK; K.Adeyeye@bath.ac.uk

* Correspondence: mohsen.besharat@tecnico.ulisboa.pt

Received: 15 June 2020; Accepted: 5 August 2020; Published: 7 August 2020

Abstract: Energy and climate change are thoroughly linked since fossil energy generation highly affects the environment, and climate change influences the renewable energy generation capacity. Hence, this study gives a new contribution to the energy generation in water infrastructures by means of an inline pumped-storage hydro (IPSH) solution. The selection of the equipment is the first step towards good results. The energy generation through decentralized micro-hydropower facilities can offer a good solution since they are independent of the hydrologic cycle associated with climate change. The current study presents the methodology and analyses to use water level difference between water tanks or reservoirs in a base pumping system (BPS) to transform it into the concept of a pump-storage hydropower solution. The investigation was developed based on an experimental facility and numerical simulations using WaterGEMS in the optimization of the system operation and for the selection of the characteristic curves, both for the pump and turbine modes. The model simulation of the integrated system was calibrated, and the conceptual IPSH that can be installed was then investigated. The achieved energy for different technical scale systems was estimated using proper dimensional analysis applied to different scaled hydraulic circuits, as well as for hydropower response.

Keywords: pumped-storage; micro-hydropower; water networks; dimensional analysis; pumping system

---

1. Introduction

The increasing need for energy in current societies is inducing more emissions of carbon dioxide to the atmosphere worsening the climate change issues. For that reason, the use of renewable energies has received serious attention in recent years. Ensuring a clean environment and a sustainable development of renewable energy sources widely and globally are appointed as future targets (in UN 2030 Agenda [1]). Although there are great interests in wind and solar as green energy sources, the hydropower should not be overlooked with huge given proof. Currently, hydropower is considered as one of the most flexible and preferred sources to produce electricity [2] and for renewable integration. Therefore, the idea of power production using water based on its available flow energy can contribute to the reduction in significant environmental impacts [3,4].

The basic principle of hydropower is driving a turbine by using the power of water through two common configurations: with or without reservoirs. In hydropower with a reservoir, the water can be stored and is able to generate a considerable amount of energy depending on its capacity, while in hydropower without a reservoir, it produces less, operating preferentially with a constant flow, such as water trunk mains or transmission lines [5]. For that reason, among the diverse use of water in multipurpose systems, one usage is increasing for generating energy [6,7]. This is a field in which large potential in the micro-hydropower (MHP) category with a low or medium head is available in different conveyance systems of water networks [8,9].

Many studies already exist in the literature, but still, there is space for further explorations. In that realm, tubular propeller turbines provide a good possibility that has been addressed in the study [10] presenting an experimental work on the characterization of an inline tubular propeller suitable for pressurized systems, such as water supply and distribution networks, reporting an efficiency around 60% for low-head (below 50 m) operations. Another study [11] has presented an investigation about the reduction in the energy consumption in water pipe networks through the use of low-cost MHPs as a means to exploit the excess pressure within these networks to produce stand-alone electricity production for local or rural consumption. In a case study, Samora et al. [12] developed a method being applied to the city of Fribourg in Switzerland and analyzed the benefits associated with a proposed hydropower scheme. The optimization led to more economic installations, and also combinations of one or more series turbines tested in this study, increasing the energy production. Additionally, a new integrated technical solution with economic and system flexibility benefits is introduced through replacing a pressure reduction valve (PRV) by a pump as turbine (PAT) [13]. In that sense, an implementation in a system having a head between 35 and 90 m reported energy generation of 20 to 94 MWh for flow rate varying between 20 and 50 L/s [14]. Choosing PATs instead of conventional turbines is a good practice that reduces the initial costs of the energy system [15,16]. Additionally, the excess energy in water networks can be recovered to optimize the energy efficiency of the systems [17] equipped with a pump station but presenting an excess of available energy in a gravity pipe branch. In order to use this excess of hydraulic available energy, a water turbine can be installed providing good energy results [17]. Very valuable studies exist on energy generation in water distribution networks (WDNs). Pérez-Sánchez et al. [18] tried to study the sustainability of WDNs by adding energy recovery possibility which leads to an added value to those systems. This study investigated different solutions in energy recovery and provided useful recommendations. Another recent study [19] introduced a novel system for WDNs to control the pressure using a hydraulically operated PRV and generate energy using a PAT. Hydropower generation in a small-scale WDN has also been investigated in another study by using PAT [20]. This study tried to characterize the PAT specification installed instead of a PRV by examining different scenarios. The literature review in WDN shows that most of the energy generation studies use PAT either as a single or hybrid solution. In that realm, Carravetta et al. [21] introduce PAT design strategies for WDNs based on a comparison between hydraulic and electrical regulations in a variable operating strategy. This study showed that the hydraulic regulation of the PAT leads to higher flexibility and efficiency. However, the flow rate is continuously changing in WDNs, which implies having a control system to avoid deficiency in the energy generation. In that case, rather than average values, the daily variation of the flow must be considered. Some studies suggest a real time control (RTC) strategy [22]. Creaco et al. recently presented a comprehensive study of different RTC methods in WDNs [23]. In that scope, Puleo et al. [24] studied the PAT application in WDNs by considering the flow variation that was driven by variable head tanks. This study used a global gradient algorithm to more accurately simulate the network parameters proving that the PAT efficiency is dependent on the network supply and pressure conditions. The flow variation subject is also studied by Alberizzi et al. [25] in a WDN in Italy when a PAT speed control strategy was applied to better exploit the flow rate even in high variations. The introduced strategy was applied and it was possible to provide energy recovery values of the order of 30% higher than a no speed variation scenario. The application of MHP solutions has gone even further to irrigation networks with a promising future of their applicability [26–30]. Additionally, novel solutions using the compressibility effect of air have been presented in some studies [9,31–33] that can be combined with pumped-storage hydropower (IPSH) to offer a hybrid pump–hydro solution. Despite the mentioned studies in this field, it is still worth exploring more aspects. Most of the mentioned solutions are completely dependent on a considerable available head to produce energy that mostly occurs through exploiting PATs. In that sense, this study tried to examine the idea of creating the required head by adding storage tanks to the system.

Hence, the current study introduces an inline pumped-storage hydropower (IPSH) solution based on experimental tests, numerical simulations and parametric analysis. Among several discussed practical solutions, IPSH can add more flexibility to pumping systems by providing higher head difference enabling the reduction in energy consumption through reusing the pumped water in gravity branches. Since it will not ask for major changes and large additional investments, using a by-pass to the main pumping system can offer a profitable hydro-energy solution. Among the energy generated, the water power potential energy is more flexible, adaptive and feasible to combine with other renewable sources to feed the pumping system as a hybrid solution. Based on that, this study aims to (a) define the electromechanical selection rules; (b) present a conceptual idea of energy generation by using storage tanks in WDN; (c) analyze results from an experimental small-scaled system by means of numerical simulations to calculate energy generation in a conceptual energy recovery prototype; (d) use suitable dimensional analysis (for hydraulic system and turbomachinery) to predict energy output in different systems' scales. In general, this study presents a low-cost energy prediction using a pumped-storage hydropower solution through experimental measurements, calibration of the numerical model and dimensional analyses.

2. Electromechanical Equipment

2.1. Pump Characteristics

2.1.1. Characteristic Curves and Operational Point

Pump characteristic curves describe the relationship between the flow rate and the pump head for a specific pump type (Figure 1). Other important information is graphs for different impeller diameters, the net positive suction head (NPSH), the efficiency and power curves. In the case of any pump, its designation determines its specific nominal discharge impeller diameter. The pump's efficiency throughout its characteristic curve should not drift too much from the best efficiency point (BEP). The motors whose pole number is associated with the rotational speed value also have their own efficiencies to be considered.

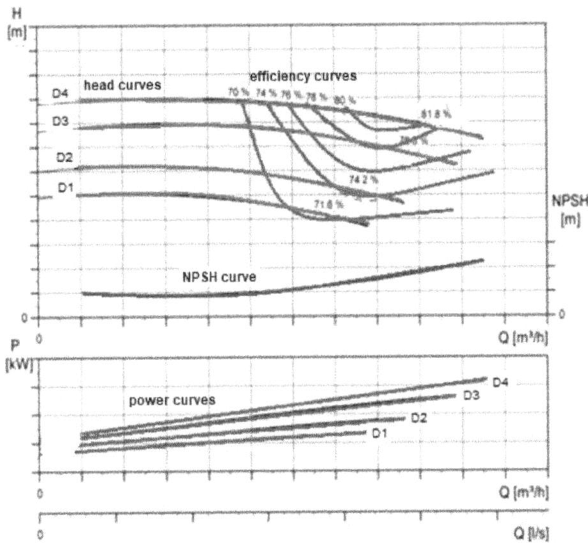

Figure 1. Typical pump characteristic curves: head, efficiency, NPSH and power curves.

The operation point in each pump curve is dependent upon the characteristics of the system in which it is operating. The system head curve is the head equation or the relationship between flow and hydraulic losses in the hydraulic system. Figure 2 shows the pump's operating point (Q_1, H_1) which can change with the differential water level, closure flow control valve or the rotational speed of the pump.

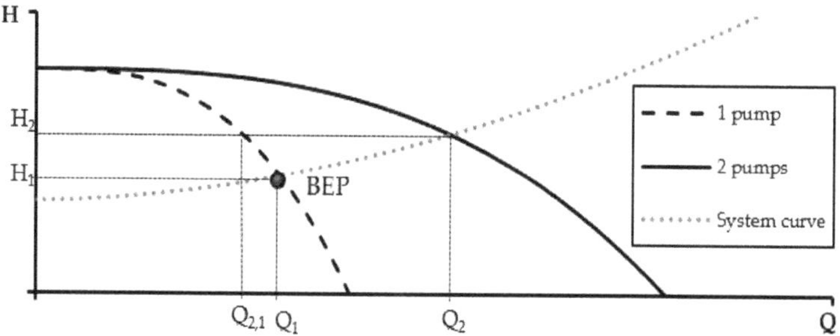

Figure 2. Representation of a single and parallel pump curves.

When two or more pumps were installed in parallel, the increasing of flow rates were obtained (Figure 2). The operating point (Q_2, H_2) represented a higher volumetric flow rate than a single pump for a consequence of greater system head loss, and the volumetric flow rate was lower than twice the flow rate achieved by using a single pump $(Q_{2,1} < Q_1)$. All of these changes can influence the system efficiency.

The affinity laws expressed in Equation (1) represent the mathematical relationship between the rotational speed (n), flow rate (Q), head (H) and pump power (P) for the same impeller diameter. The pump specific speed is given by Equation (2).

$$\frac{N_1}{N_2} = \frac{Q_1}{Q_2} = \sqrt{\frac{H_1}{H_2}} = \sqrt[3]{\frac{P_1}{P_2}} \tag{1}$$

$$n_{sp} = N \frac{Q^{1/2}}{H^{3/4}} \tag{2}$$

Variable speed drive (VSD) in pumps induces smooth speed variations of the rotating shaft, directly proportional to the flow, translating into significant pump power variations, which can increase the efficiency of the pumping operation when compared to pumps equipped with fixed speed drive (FSD). Nevertheless, this also affects the pump head, rendering it inoperable below the point where it will not cross the system curve. As possible operating points come closer the operation of the pump becomes unstable if transient regimes occur, inducing the flow variation. Additionally, the pump curve can have a shut-off head inferior to the system curve, meaning that the pump is not able to start at that particular speed.

2.1.2. Selection of a Pump

In a water network system, the total water supply needed by the population must be met by the operation of one or more pumps, whereby the flow rate must be superior to the average daily demand, considering pumps with a flow up to double that value. Regarding the total head, the pumps should comprise a range which takes into account the increase in roughness of the pipes over time such as through Hazen–Williams (H–W) roughness coefficients. Based on the available pump curve, the pump needs to be suitable for its water supply system (WSS) or it will probably operate with reduced efficiency or even with flow instabilities. To have a real idea with a pump efficiency at a BEP of 72% and a motor with 92% efficiency, the total efficiency at that point is 66%. Since the operating point is far from the BEP, the operational costs resulting from additional energy consumption can be quite significant.

Then, in a pump system design, several pumps which fitted the considered flow and head ranges can be selected, such as the practical application presented in Figure 3.

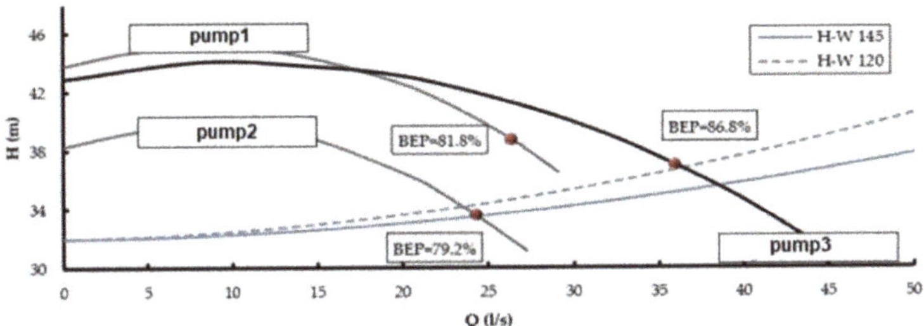

Figure 3. Different pumps and system curves for different head losses.

1. Pump 1 is not appropriate since it would operate with flow rates not recommended by the manufacturer;
2. Pump 2 has a flow rate near to the average daily demand, increasing its probability of becoming obsolete if the demand is intensified or if the flow is reduced due to a pipe roughness increase over time. The maximum efficiency is also inferior to the pump 3 and, if equipped with a VSD, the possible speed range is minor given its inferior heads.

It is also relevant to analyze the speed range in which a pump can operate, where the solution space in the optimization process by the hydraulic simulator WaterGEMS does not include unfeasible solutions, which is determined by comparison to the system curve. The selection has to avoid system instability considering all the former considerations on the pump selection.

2.2. Pump as Turbine Curves

When a pump works in the turbine zone, the motor will operate as a generator. During pump operation, the discharge, Q, is a function of the rotating speed, n, and the pumping head, H, whereas the alteration of the speed will depend upon the torque of the motor, T.

For normal turbine operation, the rotating speed (n) and discharge (Q) were negatives and the head (H) and torque (T) were positives (Figure 4).

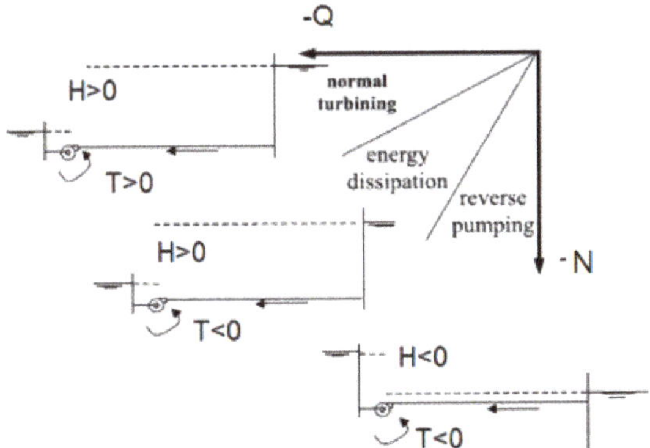

Figure 4. Pump operation zone in the third quadrant and characteristic parameters [34].

Based on pumps operating as turbines, the turbine characterization was developed by the following next steps: (i) different parameters must be defined based on an established database of synthetic PATs (a large number of different PATs) through characteristic curves that are shown in Figure 5 as well as changing the specific speed (n_{sT}) defined according to Equation (3):

$$n_{sT} = N_R \frac{P_R^{1/2}}{H_R^{3/4}} \, (in\, m, kW)$$

(3)

where N_R is the rated rotational speed in rpm, P_R is the rated power in kW, H_R is the rated head in m, since R means the rated conditions or the PAT design point for the best efficiency condition; (ii) when the specific speed is defined, n_{sT} the values of head number (ψ) for each flow rate number (φ) can be estimated. When the specific speed was defined and the n_s was not known, the values of head number (ψ_{int}) and efficiency (η_{int}) for each discharge number value (φ_{int}) were estimated by linear interpolation. When the non-dimensional number was defined, for each diameter and rotational speed (N) the head and efficiency curves were determined by Equations (4)–(6):

$$Q = \varphi_{int} N D^3$$

(4)

$$H = \frac{\psi_{int} N^2 D^2}{g}$$

(5)

$$\eta = \eta_{int}(\varphi_{int})$$

(6)

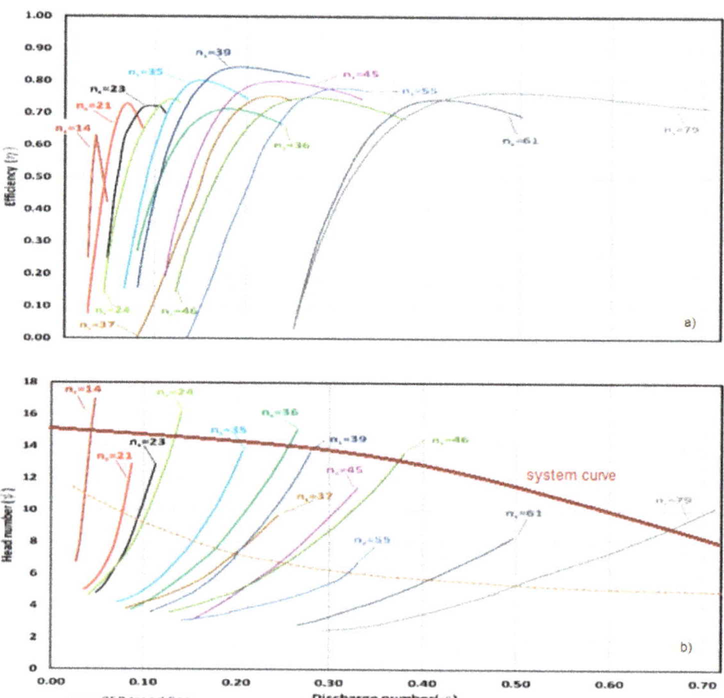

Figure 5. Pump as turbine (PAT) characteristic curves for different specific speed values, based on experimental tests and affinity laws (adapted from [25]): (**a**) efficiency curves; (**b**) head curves.

Hence, the correspondent net head and flow rate for the turbine mode will be also influenced by the system curve of the hydraulic system as represented in Figure 5.

3. Methodology

This research used a hydraulic numerical simulation model of an energy recovery system to evaluate the conceptual idea of an IPSH solution and also to estimate the potential energy available in a WSS. This approach used an experimental apparatus of a pumping system to collect the required data of pressure, flow rate, efficiency and rotational speed of a pump for a certain controlling flow. These data were exploited to establish a numerical model using a WaterGEMS simulation tool and also to calibrate the model for different operating conditions. When the numerical model for the pumping system was validated to reproduce the measured data, it was upgraded by adding a by-pass branch to create the desired energy recovery system. The numerical results from the model were used to estimate the energy output of a possible energy recovery solution. This approach has been depicted in the current section by, first, discussing the experimental system and presenting measurement data. The experimental system is known as the base pumping system (BPS) since it is the base of the future energy recovery system. Then a discussion about the calibration of the numerical model will be provided. This section will be closed by introducing the energy recovery system known as inline pumped-storage hydropower (IPSH) through numerical modelling in a WaterGEMS environment.

3.1. Base Pumping System (BPS) and Experimental Results

3.1.1. System Configuration

The experimental facility BPS is located at the laboratory of hydraulic (LH), Instituto Superior Técnico, Universidade de Lisboa and consists of several components (Figure 6): (a) a centrifugal pump that feeds the loop pipe system; (b) an interface control panel; (c) an electromagnetic flow meter to measure the instant flow (SC-1); (d) flow control valves along the pipe at the inlet and outlet of the pumps (VR-1 and VR-2); (e) pressure transducers (SP-1 and SP-2) to record the pressures; (f) a free-surface reservoir (water tank). Pipes are made of polyethylene with a diameter of 25 mm and a total length of 198 cm.

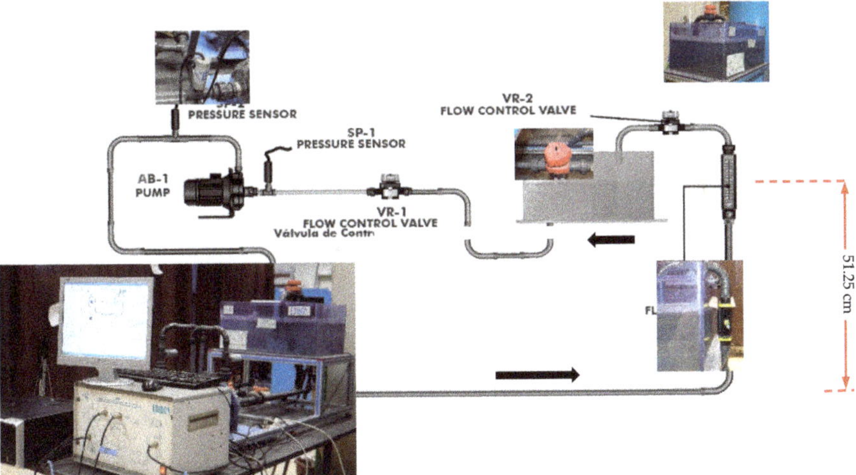

Figure 6. Experimental facility at Instituto Superior Técnico (IST) lab.

The measuring range of the mentioned sensors is presented in Table 1. A three-phase motor activated the pump, while the interface control panel provided the possibility of adjustment and measurement of the rotating speed and also the transmitted mechanic torque.

Table 1. Measuring range of test parameters.

Parameters	Measuring Range
Pump head [m]	1.8 to 16.92
Flowmeter Rate [L/min]	0 to 61.98
Rotational speed of the pump [rpm]	1600 to 2950
Opening valve VR2 [%]	0 to 100

A free surface tank with a capacity of 85 dm^3 existed at downstream of the operating system. A valve located at downstream made it possible to induce flow variations by maneuvering and applying local head loss of the valve. An electromagnetic flowmeter (SC-1) and two pressure transducers (SP-1 and SP-2) were used to record the flow rate and pressure data, respectively. Different tests were carried out for different conditions depending on the different opening percentages of the VR-2 valve and also the pump rotational speeds, as presented in Table 2. The data measurements included the flow rate (Q), head (H), rotational speed (N), pump upstream and downstream pressures, the efficiency of the pump (η) and hydraulic and mechanic powers P_h and P_M, respectively).

Table 2. Experimental tests specifications.

Variable Parameters	Tested Values
VR-2 closure percentage [%]	4.16, 6.25, 8.33, 10.41, 12.5, 16.66, 25, 33.33, 50, 66.60, 83.33, 100
Pump rotational speed, N [rpm]	1600, 1800, 2000, 2200, 2400, 2600, 2800, 2950

3.1.2. Experimental Results

During the pump operation, the flow rate was a function of the rotational speed and the pumping head [35]. The flow rate varied from zero, for a fully closed valve, to 61.98 L/min for a rotational speed of 2950 rpm, as shown in Figure 7. Additionally, the minimum and maximum measured heads were 1.8 and 16.92 m for rotational speeds of 1600 and 2950 rpm, respectively. Hence, Figure 7 presents characteristic curves for different rotational speeds, flow rate and head, covering the range of operation for the pumped storage system and efficiency variation for N = 2950 rpm.

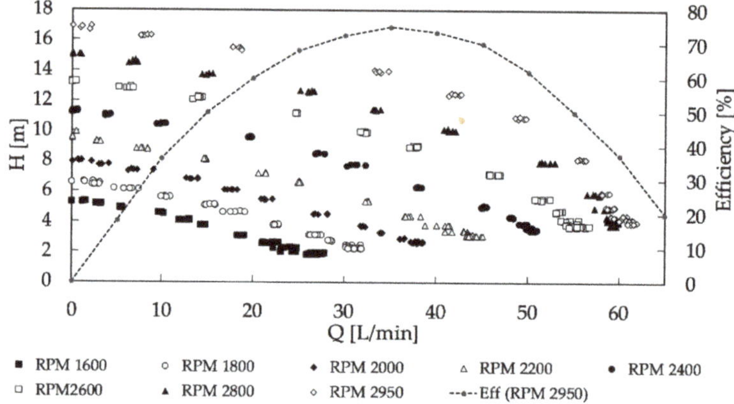

Figure 7. Characteristic curves obtained from the experimental tests.

3.2. Model Calibration

To perform the system analyses, it was necessary to calibrate the numerical model based on the measured data. The WaterGEMS software from Bentley was used for numerical simulation providing an optimized simulation tool and a user-friendly environment for water distribution networks. WaterGEMS calculated the hydraulic head and pressure at every node along with the flow rate, the flow velocity and the head loss in each pipe branch and as well as the hydraulic head using the gradient algorithm based on the EPANET solver. The water system shown in Figure 8 was built by including one tank, one centrifugal pump and two flow control valves similar to the experimental setup. The experimental setup (shown in Figure 6) is called BPS (base pumping system) in this paper.

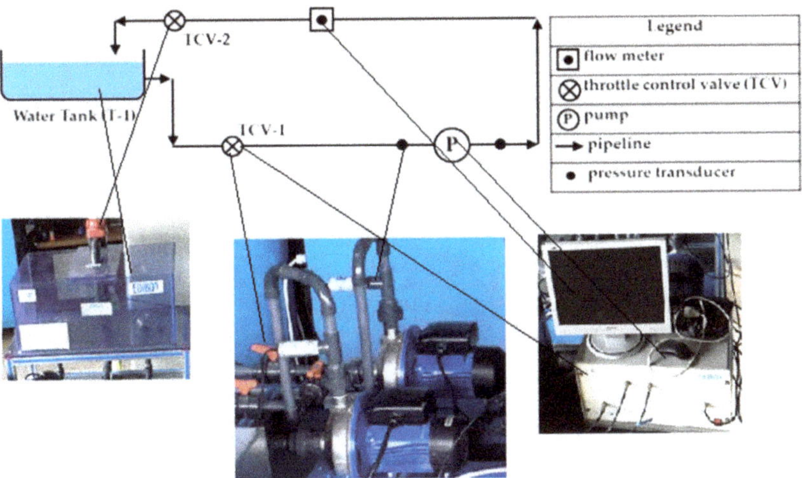

Figure 8. Scheme of the base pumping system (BPS) in the mathematical model and experimental apparatus.

The calibration process involved the optimization of BPS parameters correspondent to actual measured conditions [35,36]. The BPS was calibrated considering the characteristic curves of the pump, for several pump rotational speeds and valve opening percentages (TCV-2).

Hence, in the simulation process, a variable speed pump (as a VSD) was introduced into WaterGEMS through proper pump curves. The rated pump characteristic curve was defined for the maximum rotational speed, i.e., 2950 rpm, and for each different rotational speed a relative speed factor (RSF) was defined as a coefficient of this maximum rotational speed. A throttle control valve was also defined to induce different flow rates into the system by the partial opening of the valve. In summary, each test was simulated by defining proper RSF and then changing opening flow percentages as in Table 2. This process was repeated for different RSFs until all the rotational speeds were simulated. During this simulation process, the valve discharge coefficients and other associated losses were optimized and calibrated. Results obtained from the simulation by WaterGEMS presented in Figure 9a show good relative accordance with the measured data, associated with scale effects, offering the root mean square errors (RMSEs) shown in Figure 9b with an average of around 0.72. The numerical model calibration based on experimental data guaranteed reliability to follow simulations that take place to assess the behavior of a new adaptation system for energy recovery.

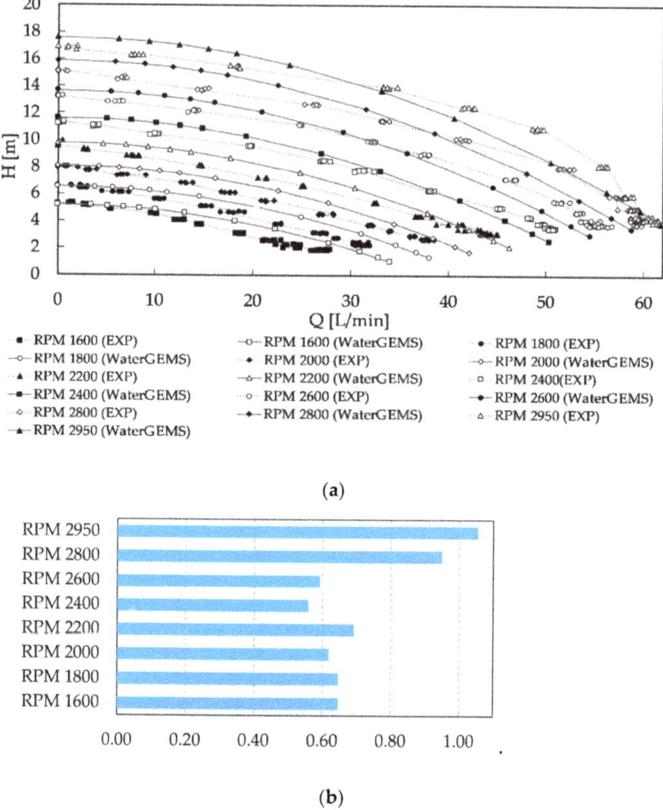

Figure 9. Comparison of experimental and numerical results for the BPS; (**a**) head and flow rate; (**b**) root mean square error (RMSE).

A traditional pumping system is composed of different elements that correspond to energy consumption and head losses. However, interesting potential usually exists in pipe branches of pumping systems for energy recovery which can supply energy to treatment plants, electric data base measurement and control devices, and in general, reduce costs of energy in water networks.

3.3. Pumping System Operation

The best efficiency point (Q_R, H_R, η_R) of the pump was characterized by a flow rate of 35 L/min, with a head of 13.5 m and an efficiency of 75% for the rotational speed of 2950 rpm and specific speed 10.05 (Table 3).

Table 3. Best efficiency point in the pumping system.

Parameter	Value
Flow rate [L/min]	35
Head [m]	13.5
Efficiency [%]	75
Rotational speed [rpm]	2950
Specific speed (Equation (2))	10.05

The characteristic curves in the dimensionless form (Figure 10) were constructed based on a rated condition associated with the best efficiency point. The dimensionless curves provided a tool to transfer information to other equivalent systems.

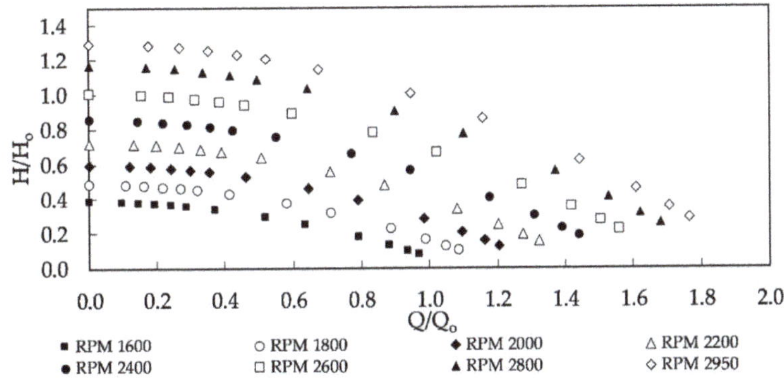

Figure 10. Dimensionless pump characteristic curves for different rotational speeds.

The best efficiency point of the pump was selected to present the head and efficiency variations with a flow rate, for the rated rotational speed of 2950 rpm, in Figure 11.

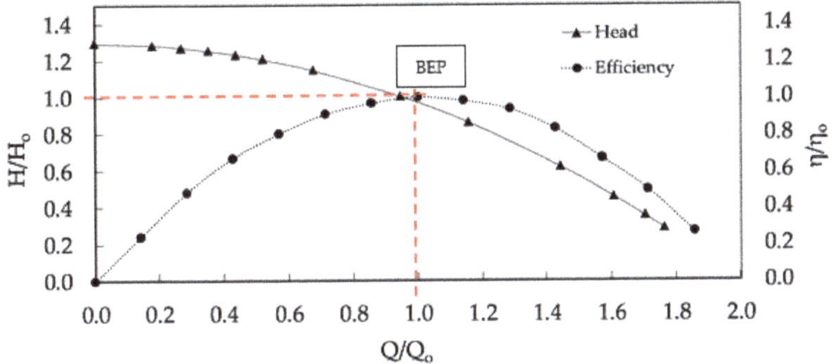

Figure 11. Dimensionless pump curves for the rotational speed of 2950 rpm.

3.4. Inline Pumped-Storage Hydropower (IPSH)

In this stage, the numerical model was ready to be upgraded to an energy recovery system. The improvement was performed by adding a by-pass line, as shown in Figure 12. In order to examine the IPSH solution, a free surface tank (T-2) was installed in the by-pass branch that was added to the base pumping system (BPS) with the capacity of 0.032 m^3. The by-pass line was equipped with a throttle control valve (TCV) working in an open or closed position to include or isolate the by-pass line. The T-2 was located at a lower elevation to generate a gravity flow from T-1 to T-2. The new IPSH system was considered as a loop system to use the previously assessed characteristics attained in the experimental loop system. It is worth mentioning that the idea is not limited to loop systems but can be adapted to a real system with direct flow condition. In a direct flow system, based on the available head at T-1 and downstream demand, the by-pass line can be activated to use the available head difference for energy generation. Hence, a turbine was considered in the by-pass line to generate energy from the gravity flow. Since WaterGEMS does not include a built-in turbine element, the general purpose valve (GPV) was used for this purpose by defining the flow-head loss curve correspondent to the turbine characteristic curves.

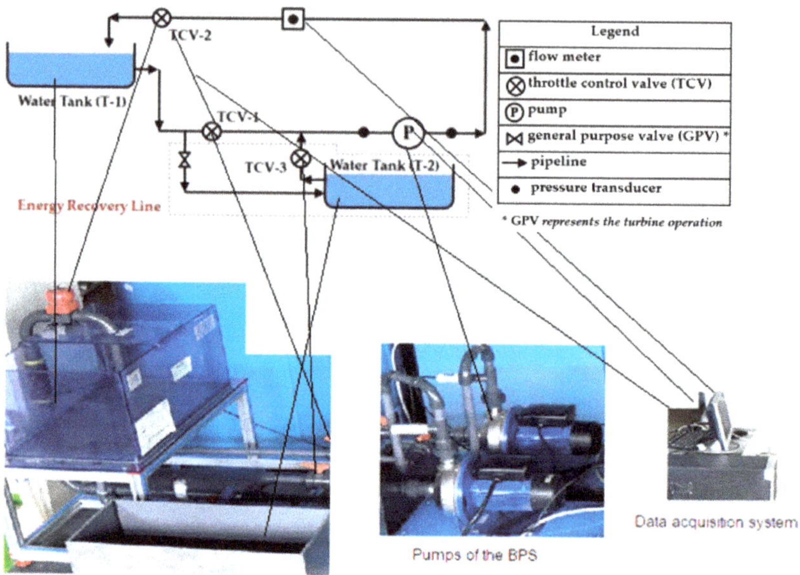

Figure 12. Scheme and visualization of the experimental set-up of the inline pumped-storage hydro (IPSH) solution.

The numerical simulation was used to assess the main variations in the IPSH system. The characteristic curves in Figure 13 were used to adjust the flow rate in the main pipe branch by changing the rotational speed of the pump and the TCV-2 opening based on the flow rate in gravity by-pass line for energy recovery. The simulations were carried out for an extended period of 24 h. Two scenarios were considered, i.e., identical and variable flow rates in the pipe system. If the flow rates in different branches of the system were equal, it led to a steady flow regime resulting in a constant water level in T-1 and T-2 (Figure 13). In this case, a gravity flow rate of 29.11 L/min existed from T-1 to T-2. To maintain this flow rate in the main pipe, a pump rotational speed of 2600 rpm and a TCV-2 opening of 72.5% were found to be appropriate based on the experimental measurements. In other words, the characteristic curves gave the ability to accurately adjust the pump working condition and valve opening percentage in order to establish a flow rate equal to the gravity flow between T-1 and T-2. Based on that, the water levels in T-1 and T-2 for the identical flow rate scenario over 24 h remained constant, as 0.50 and 0.18 m, respectively.

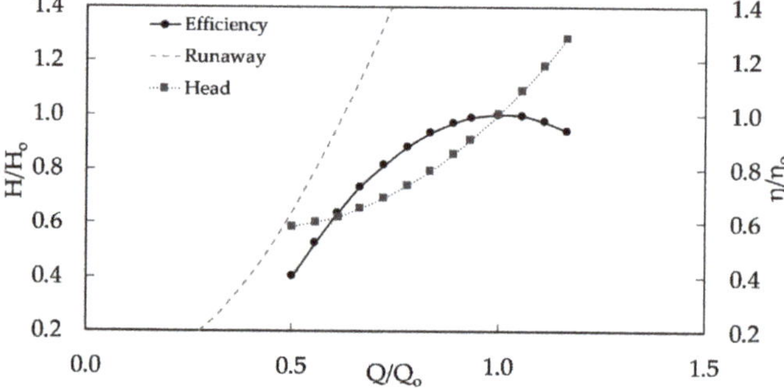

Figure 13. Dimensionless turbine characteristic curves based on numerical simulations.

In a water supply system, there was a pattern of flow demand along each 24 h. This pattern had a typical representation of each system characterization depending on the flow consumption used for water network design. Therefore, in this research, the same procedure was adopted since the flow pattern varied along the time, between rush consumption or peak hours to fewer consumptions, normally associated with the night period—the one which was also used for leak detection, since the level of consumption attained the minimum. Hence, a suitable design project for this extended period of 24 h represents a more granted solution to face flow and water level variations, head losses, leakages occurrence associated with high pressure values, along the hydraulic system and machine operation adaptation, which fit both results as the operating point. This is a complex issue that requires an extended period and is typically used in the design of water systems. Under some operating conditions, the flow rate in different branches can be variable and unequal. Then, to evaluate this scenario, a flow rate pattern of 24 h was adapted, based on a typical demand configuration, as presented in Figure 14a. A TCV-2 valve closure pattern was then calibrated to induce the desired flow rate in the hydraulic system (Figure 14b).

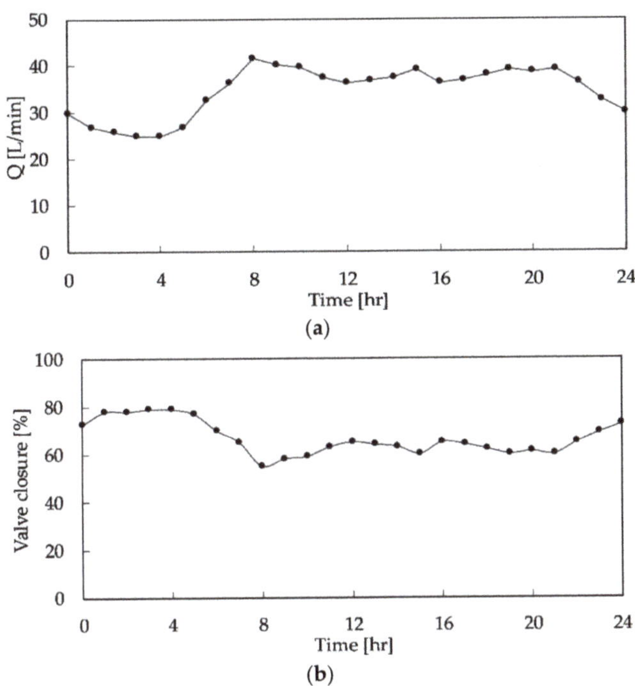

Figure 14. A real system pattern: (**a**) flow rate pattern; (**b**) valve closure pattern.

Despite the previous case of having a balance of flow rate in the whole system and constant water levels in tanks, the water level in both tanks changed with time, as shown in Figure 15. The water level variations in T-1 and T-2 were correlated, decreasing in T-1, in turbine mode and increasing in T-2, in pump mode, in a controlled optimized way between the maximum and minimum limits for tank water levels (Figure 15).

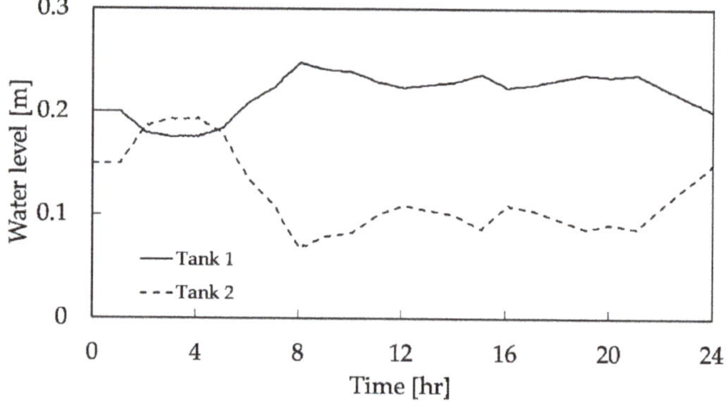

Figure 15. The water level in T-1 and T-2 based on the variable demand pattern.

The operating curve of the GPV valve (acting as a turbine) was calculated with the available gross head and total head losses to obtain the net head of the turbine, as presented in the dimensionless graph of Figure 13. Equation (7) was used to calculate turbine power:

$$P = \gamma Q H \eta \tag{7}$$

where P is the power, γ is the specific weight of the water, Q is the flow rate, H is the turbine head and η is the efficiency. Equation (7) was exploited to calculate the power for two different mentioned scenarios of constant and variable flow rates, as presented in Figure 16. The energy production of the system for a fixed flow rate of 35 L/min was 1.24 kWh, while the variable scenario led to 1.81 kWh daily energy production.

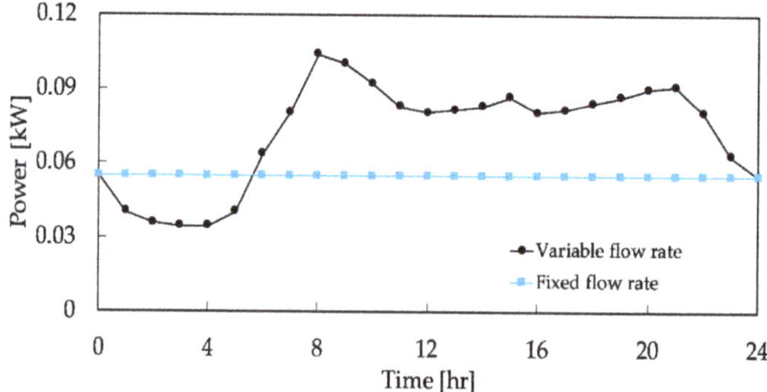

Figure 16. Power in the fixed and variable demand patterns.

4. Dimensional Analysis and Discussion

Experimental tests are used to predict the performance of a real device including the behavior under different operating conditions. In this case, the model was evaluated on a laboratory scale, and then the results were simulated for real conditions. The similarity theory requires principles of geometric, kinetic and dynamic parities between a model and a prototype. These affinity laws in different categories can be attained by expressing shape, size, velocity component and acting forces [37].

In this study, the scales of velocity and flow rate were calculated based on the Froude criterion, as analyzed in [38], from the length scale (λ_L) using the following equations:

$$\lambda_L = \frac{L_{mod}}{L_{pro}} \tag{8}$$

$$\text{Flow rate scale}: \; Q = V.A \Rightarrow \lambda_Q = \lambda_V.\lambda_L^2 \Rightarrow \lambda_Q = \lambda_L^{5/2}, \tag{9}$$

$$\text{Velocity scale}: \; \frac{V_{mod}}{V_{pro}} = \frac{V_{mod}}{(gL_{mod})^{1/2}} = \frac{V_{pro}}{(gL_{pro})^{1/2}} \Rightarrow \frac{V_{mod}}{V_{pro}} = \frac{(gL_{mod})^{1/2}}{(gL_{pro})^{1/2}} \Rightarrow \lambda_V = \lambda_L^{1/2}. \tag{10}$$

Classic similarity laws for pumps and turbines with the same impeller diameter state that the discharge is proportional to the rotational speed, while the head is proportional to the squared rotational speed as Equation (1). Then based on [39,40], the correlation between pump and turbine mode was proved to be:

$$n_{S_T} \; (\text{in } m, \, m^3/s) = 0.8793 n_{S_P} \tag{11}$$

$$\frac{Q_T}{Q_P} = 1.3595 \frac{N_T}{N_P} \tag{12}$$

To evaluate the results of the model in a technical approach, for a loop system where $Q_T = Q_P$ the length scales of 20 and 50 were considered. Additionally, for geometrically similar impellers operating at the same specific speed, the affinity laws are as follows:

$$\frac{N_{mod}}{N_{pro}} = \left(\frac{H_{mod}}{H_{pro}}\right)^{1/2} \frac{D_{pro}}{D_{mod}} = \left(\frac{H_{mod}}{H_{pro}}\right)^{3/4} \left(\frac{Q_{pro}}{Q_{mod}}\right)^{1/2} = \left(\frac{P_{pro}}{P_{mod}}\right)^{1/2} \left(\frac{H_{mod}}{H_{pro}}\right)^{5/4} \tag{13}$$

or separating by specific flow, head and power:

$$\begin{aligned}
\frac{Q_{pro}}{Q_{mod}} &= \left(\frac{N_{pro}}{N_{mod}}\right)\left(\frac{D_{pro}}{D_{mod}}\right)^3 \\
\frac{H_{pro}}{H_{mod}} &= \left(\frac{N_{pro}}{N_{mod}}\right)^2\left(\frac{D_{pro}}{D_{mod}}\right)^2 \\
\frac{P_{pro}}{P_{mod}} &= \left(\frac{N_{pro}}{N_{mod}}\right)^3\left(\frac{D_{pro}}{D_{mod}}\right)^5
\end{aligned} \tag{14}$$

The results are presented in Table 4. The power of the model was 0.052 kW but, by upscaling, it grew to values of 1532 and 41,657 kW, for scales of 20 and 50, respectively.

Table 4. Scale-up parameters for a hydraulic system and turbomachine affinity characteristic parameters with n_{sT} (in m, m^3/s) = 8.8.

Parameter Unit	Hydraulic System		Turbine Impeller	Turbine Affinity Laws		
	Q	V	D	N_T	H_T	P_T
	[m^3/s]	[m/s]	[mm]	[rpm]	[m]	[kW]
Model	0.60×10^{-3}	0.99	25	2170	10.64	0.052
1/20	1.04	4.42	500	470	200	1532
1/50	10.31	6.99	1250	298	503	41,657

5. Conclusions

In a base pumping system (BPS), the characteristic curves in pump and turbine modes were defined and analyzed in order to obtain the best operating conditions in water networks. The research study included experimental analyses, hydraulic simulations and optimized conditions to better define the best operating point. The experimental results were exploited to calibrate a numerical hydraulic simulator model in the WaterGEMS environment, which uses optimization in searching for the best operating conditions. This numerical model was then used to evaluate a smart conceptual energy recovery solution of an inline pumped-storage hydropower (IPSH) system. The characteristic parameters (i.e., flow rate, velocity, impeller diameter, rotational speed, head and power) of this novel model were calculated to provide a background for the energy estimation. As a result, the following main conclusions can be pointed out:

1. Characteristic curves of turbomachines in pump and turbine mode were defined for the best selection which conducts the best energy solution, avoiding eventual induced operating instabilities.
2. An inline pumped-storage hydropower (IPSH) solution was defined and adapted from a base pumping system (BPS) in some existing water infrastructures of small to large scales, while not requiring significant changes and investments based on a by-pass and a lower tank upstream the pumping station.
3. The energy generation using the gravitational flow appears to be an economic advantage in the definition of the energy recovery solution, as demonstrated through the achieved power.
4. Depending on the type of demand (i.e., a constant flow between tanks or a variable demand pattern with water level compensation), the application shows a smart pressure and flow control in an energy recovery solution, replacing classical flow control valves.
5. Based on similarity laws for the hydraulic system and the turbomachinery between pumps and turbines, a scaling-up approach for larger hydro energy converters was developed showing promising results.
6. The smart approach based on a controlled recovery energy solution, which would be dissipated and the increasing of the system flexibility by a new bottom tank allowing two types of flow conditions (i.e., pump and turbine modes), can significantly improve the energy efficiency in the water sector, allowing us to better face the associated existent energy costs (e.g., pumping, treatment plants, water leakage, expansion and reparation of infrastructures and water bill for costumers).

Author Contributions: Conceptualization—H.M.R.; Methodology and formal analysis—H.M.R., A.D., M.B.; Writing, review & editing—H.M.R., A.D., M.B. and K.A.; Data collection and validation—A.D. and M.B.; Supervision—H.M.R. and K.A.; Funding acquisition—H.M.R. and K.A. All authors have read and agreed to the published version of the manuscript.

Funding: This work was supported by funding from REDAWN European project (www.redawn.eu).

Acknowledgments: The authors would like to thank CERIS (Civil Engineering, Research, and Innovation for Sustainability) Centre from Instituto Superior Técnico, Universidade de Lisboa, Portugal for providing the experimental facilities and also the European project REDAWN (Reducing Energy Dependency in Atlantic Area Water Networks) EAPA_198/2016 from the INTERREG ATLANTIC AREA PROGRAMME 2014–2020 for the grants support.

Conflicts of Interest: The authors declare no conflict of interest.

Nomenclature

Symbols

D	turbomachinery diameter [m]
H	head [m]
L	length [m]
n	rotational speed [rpm]
n_{sp}	specific speed of the pump
n_{sT}	specific speed of the turbine
P	power [kW]
Q	flow rate [L/min] or [m^3/s]
t	time [s]
V	flow velocity [m/s]

Indices

h	hydraulic
int	interpolated value
mod	model
M	mechanical
pro	prototype
R	rated or best efficiency point
T	related to turbine

Greek letters

γ	specific weight [N/m^3]
η	efficiency
ψ	head number
φ	flow rate number

Abbreviations

BPS	base pumping system
IPSH	inline pumped-storage hydropower
MHP	micro-hydropower
PAT	pump as turbine
WDN	water distribution network
WSS	water supply system

References

1. UN General Assembly. Transforming Our World: The 2030 Agenda for Sustainable Development. 21 October 2015. A/RES/70/1. Available online: refworld.org/docid/57b6e3e44.html (accessed on 14 February 2020).
2. International Energy Agency (IEA). Renewables: Market Analysis and Forecast from 2019 to 2024, Paris. 2019. Available online: iea.org/reports/renewables-2019 (accessed on 13 February 2020).
3. Ramos, J.S.; Ramos, H.M. Solar powered pumps to supply water for rural or isolated zones: A case study. *Energy Sustain. Dev.* **2009**, *13*, 151–158. [CrossRef]
4. Ramos, J.S.; Ramos, H.M. Sustainable application of renewable sources in water pumping systems: Optimised energy system configuration. *Energy Policy* **2009**, *37*, 633–643. [CrossRef]
5. Hoes, O.A.C.; Meijer, L.J.J.; Van der Ent, R.J.; Van de Giesen, N.C. Systematic high-resolution assessment of global hydropower potential. *PLoS ONE* **2017**, *12*. [CrossRef]
6. Dadfar, A.; Besharat, M.; Ramos, H.M. Storage ponds application for flood control, hydropower generation and water supply. *Int. Rev. Civ. Eng.* **2019**, *10*. [CrossRef]
7. Kougias, I.; Aggidis, G.; Avellan, F.; Deniz, S.; Lundin, U.; Moro, A.; Muntean, S.; Novara, D.; Pérez-Díaz, J.I.; Quaranta, E.; et al. Analysis of emerging technologies in the hydropower sector. *Renew. Sustain. Energy Rev.* **2019**, *113*. [CrossRef]
8. Besharat, M.; Dadfar, A.; Viseu, M.T.; Brunone, B.; Ramos, H.M. Transient-flow induced compressed air energy storage (TI-CAES) system towards new energy concept. *Water* **2020**, *12*, 601. [CrossRef]
9. Ramos, H.M.; Zilhao, M.; López-Jiménez, P.A.; Pérez-Sánchez, M. Sustainable water-energy nexus in the optimization of the BBC golf-course using renewable energies. *Urban Water J.* **2019**, *16*, 215–224. [CrossRef]
10. Samora, I.; Hasmatuchi, V.; Münch-Allign, C.; Franca, M.J.; Schleiss, A.J.; Ramos, H.M. Experimental characterization of a five blade tubular propeller turbine for pipe inline installation. *Renew. Energy* **2016**, *95*, 36–366. [CrossRef]
11. Carravetta, A.; Fecarotta, O.; Ramos, H.M.; Mello, M.; Rodriguez-Diaz, J.A.; Morillo, J.G.; Kemi Adeyeye, K.; Coughlan, P.; Gallagher, J.; McNabola, A. Reducing the Energy Dependency of Water Networks in Irrigation, Public Drinking Water, and Process Industry: REDAWN Project. *Proceedings* **2018**, *2*, 681. [CrossRef]
12. Samora, I.; Manso, P.; Franca, M.J.; Schleiss, A.J.; Ramos, H.M. Energy Recovery Using Micro-Hydropower Technology in Water Supply Systems: The Case Study of the City of Fribourg. *Water* **2016**, *8*, 344. [CrossRef]
13. Fecarotta, O.; Aricò, C.; Carravetta, A.; Martino, R.; Ramos, H.M. Hydropower Potential in Water Distribution Networks: Pressure Control by PATs. *Water Resour. Manag.* **2014**, *29*, 699–714. [CrossRef]

14. Gallagher, J.; Harris, I.M.; Packwood, A.J.; McNabola, A.; Williams, A.P. Strategic assessment of energy recovery sites in the water industry for UK and Ireland: Setting technical and economic constraints through spatial mapping. *Renew. Energy* **2015**, *81*, 808–815. [CrossRef]

15. Qian, Z.; Wang, F.; Guo, Z.; Lu, J. Performance evaluation of an axial-flow pump with adjustable guide vanes in turbine mode. *Renew. Energy* **2016**, *99*, 1146–1152. [CrossRef]

16. Carravetta, A.; Derakhshan Houreh, S.; Ramos, H.M. *Pumps as Turbines*; Springer Tracts in Mechanical Engineering; Springer: Cham, Switzerland, 2018.

17. Vieira, F.; Helena, M.; Ramos, H.M. Optimization of operational planning for wind/hydro hybrid water supply systems. *Renew. Energy* **2009**, *34*, 928–936. [CrossRef]

18. Pérez-Sánchez, M.; Sánchez-Romero, F.J.; Ramos, H.M.; López-Jiménez, P.A. Energy Recovery in Existing Water Networks: Towards Greater Sustainability. *Water* **2017**, *9*, 97. [CrossRef]

19. Fontana, N.; Giugni, M.; Glielmo, L.; Marini, G.; Raffaele, Z. Use of hydraulically operated PRVs for pressure regulation and power generation in water distribution networks. *J. Water Resour. Plann. Manag.* **2020**, *146*. [CrossRef]

20. Postacchini, M.; Darvini, G.; Finizio, F.; Pelagalli, L.; Soldini, L.; Di Giuseppe, E. Hydropower Generation Through Pump as Turbine: Experimental Study and Potential Application to Small-Scale WDN. *Water* **2020**, *12*, 958. [CrossRef]

21. Carravetta, A.; Del Giudice, G.; Fecarotta, O.; Ramos, H.M. PAT Design Strategy for Energy Recovery in Water Distribution Networks by Electrical Regulation. *Energies* **2013**, *6*, 411–424. [CrossRef]

22. Fontana, N.; Giugni, M.; Glielmo, L.; Marini, G.; Verrilli, F. Real time control of a PRV in water distribution networks for pressure regulation: Theoretical framework and laboratory experiments. *J. Water Resour. Plann. Manag.* **2018**, *144*. [CrossRef]

23. Creaco, E.; Campisano, A.; Fontana, N.; Marini, G.; Page, P.R.; Walski, T. Real time control of water distribution networks: A state-of-the-art review. *Water Res.* **2019**, *161*. [CrossRef]

24. Puleo, V.; Fontanazza, C.M.; Notaro, V.; De Marchis, M.; Freni, G.; La Loggia, G. Pumps as turbines (PATs) in water distribution networks affected by intermittent service. *J. Hydroinform.* **2013**. [CrossRef]

25. Alberizzi, J.C.; Renzi, M.; Righetti, M.; Pisaturo, G.R.; Rossi, M. Speed and Pressure Controls of Pumps-as-Turbines Installed in Branch of Water-Distribution Network Subjected to Highly Variable Flow Rates. *Energies* **2019**, *12*, 4738. [CrossRef]

26. Chacón, M.C.; Rodríguez Díaz, J.A.; Morillo, J.G.; McNabola, A. Hydropower energy recovery in irrigation networks: Validation of a methodology for flow prediction and pump as turbine selection. *Renew. Energy* **2020**, *147*, 1728–1738. [CrossRef]

27. Pérez-Sánchez, M.; Sánchez-Romero, F.J.; López-Jiménez, P.A.; Ramos, H.M. PATs selection towards sustainability in irrigation networks: Simulated annealing as a water management tool. *Renew. Energy* **2018**, *116*, 234–249. [CrossRef]

28. Morillo, J.G.; McNabola, A.; Camacho, E.; Montesinos, P.; Rodríguez Díaz, J.A. Hydro-power energy recovery in pressurized irrigation networks: A case study of an Irrigation District in the South of Spain. *Agric. Water Manag.* **2018**, *204*, 17–27. [CrossRef]

29. Pérez-Sánchez, M.; Sánchez-Romero, F.J.; Ramos, H.M.; López-Jiménez, P.A. Modeling Irrigation Networks for the Quantification of Potential Energy Recovering: A Case Study. *Water* **2016**, *8*, 234. [CrossRef]

30. Pérez-Sánchez, M.; Sánchez-Romero, A.J.; Ramos, H.M.; López-Jiménez, P.A. Optimization Strategy for Improving the Energy Efficiency of Irrigation Systems by Micro Hydropower: Practical Application. *Water* **2017**, *9*, 799. [CrossRef]

31. Besharat, M.; Tarinejad, R.; Aalami, M.T.; Ramos, H.M. Study of a compressed air vessel for controlling the pressure surge in water networks: CFD and experimental analysis. *Water Resour. Manag.* **2016**, *30*, 2687–2702. [CrossRef]

32. Pottie, D.L.F.; Ferreira, R.A.M.; Maia, T.A.C.; Porto, M.P. An alternative sequence of operation for Pumped-Hydro Compressed Air Energy Storage (PH-CAES) systems. *Energy* **2019**. [CrossRef]

33. Odukomaiya, A.; Abu-Heiba, A.; Graham, S.; Momen, A.M. Experimental and analytical evaluation of a hydro-pneumatic compressed-air Ground-Level Integrated Diverse Energy Storage (GLIDES) system. *Appl. Energy* **2018**, *221*, 75–85. [CrossRef]

34. Ramos, H.; Borga, A. Pumps as turbines: An unconventional solution to energy production. *Urban Water* **1999**, *1*, 261–263. [CrossRef]

35. Ramos, H.; Borga, A. Pumps yielding power. *Dam Eng. Water Power Dam Constr.* **2000**, *10*, 197–217.
36. Fontanella, S.; Fecarotta, O.; Molino, B.; Cozzolino, L.; Della Morte, R. A Performance Prediction Model for Pumps as Turbines (PATs). *Water* **2020**, *12*, 1175. [CrossRef]
37. Chaker, M.A.; Triki, A. Investigating the branching redesign strategy for surge control in pressurized steel piping systems. *Int. J. Pres. Ves. Pip.* **2020**, *180*, 104044. [CrossRef]
38. Kapelan, Z. Calibration of Water Distribution System Hydraulic Models. Ph.D. Thesis, University of Exeter, Exeter, UK, 2010.
39. Wylie, E.; Streeter, V.; Suo, L.F. *Fluid Transient in Systems*; Prentice-Hall: Englewood, NJ, USA, 1993.
40. Ramos, H.M. Simulação e Controlo de Transitórios Hidráulicos em Pequenos Aproveitamentos Hidroelétricos. Ph.D. Thesis, Civil Engineering, Instituto Superior Técnico, Universidade de Lisboa, Lisboa, Portugal, 1995. (In Portuguese)

© 2020 by the authors. Licensee MDPI, Basel, Switzerland. This article is an open access article distributed under the terms and conditions of the Creative Commons Attribution (CC BY) license (http://creativecommons.org/licenses/by/4.0/).

MDPI
St. Alban-Anlage 66
4052 Basel
Switzerland
Tel. +41 61 683 77 34
Fax +41 61 302 89 18
www.mdpi.com

Water Editorial Office
E-mail: water@mdpi.com
www.mdpi.com/journal/water